RESEARCH AND DESIGN OF SNOW HYDROLOGY SENSORS AND INSTRUMENTATION

Selected Research Papers

R. Attri Instrumentation Design Series (Snow Hydrology)

Dr. Raman K. Attri

ISBN: 978-981-11-9763-5 (e-book)
ISBN: 978-981-14-0342-2 (paperback)

First published: 2005
Revised: 2018
Edition: 2nd
Lead author: Raman K. Attri
Published by Speed To Proficiency Research: S2Pro©
Published at Singapore
Printed in United States of America

National Library Board, Singapore Cataloguing in Publication Data

Names: Attri, Raman K., 1973-
Title: Research and design of snow hydrology sensors and instrumentation : selected research papers / Dr
    Raman K. Attri.
Description: 2nd edition. | Singapore : Speed To Proficiency Research, [2018] | Series: R. Attri
    instrumentation design series (Snow hydrology) | Includes bibliographic references.
Identifiers: OCN 1066231513 | ISBN 978-981-14-0342-2 (paperback) | ISBN 978-981-11-9763-5 (e-book)
Subjects: LCSH: Snow--Measurement--Instruments--Design and construction. | Hydrology--Instruments--
    Design and construction. | Snow--Measurement--Remote sensing. | Hydrology--Remote sensing.
Classification: DDC 551.57840287--dc23

Speed To Proficiency

Speed To Proficiency Research: S2Pro©
A research and consulting forum
Singapore 560463
https://www.speedtoproficiency.com
rkattri@speedtoproficiency.com

R. Attri Instrumentation Design Series (Snow Hydrology)

To my beloved father, Ram Krishan.

For his relentless and tireless hard work and thousands of his sleepless nights to see me become a scientist one day.

CONTENTS

ABOUT THE BOOK

This book is a collection of eight in-depth and detailed research papers authored by Dr. Raman K Attri between 1996 to 2005. The book presents early-career scientific work by the author as a scientist at a research organization. The book provides the conceptual background and key electronics and mechanical design principles used in designing sensors and instrumentation systems to measure snow hydrological parameters. The systems discussed in this book can be used to measure snow depth, layer temperature, temperature distribution profile, surface porosity, etc. The snow parameters measured from instruments and sensors discussed in this book are integrated into larger systems and are used in computer-driven models for snow avalanche predictions.

The book presents the design challenges and design methods from electronics and instrumentation design point of view. While the book provides essential understanding of analog electronics design and associated mechanical design for snow hydrological sensors, the book also presents the background theoretical and mathematical models from snow hydrology physics that governs this electronics design.

The first research paper discusses the design control techniques used to the design a remote surface detector to detect objects with porous, uneven, irregular surfaces like snow using ultrasonic beams.

The second research paper describes signal processing techniques and electronics design approaches to design a snow depth sensor with improved sensitivity and directional response using Ultrasonic Pulse-Transit Method.

The third research paper explains theoretical and mathematical model that governs the physical, mechanical, and electronics design to implement the theory of Arrayed Ultrasonic transducers to shape up the directional response and beam width of an ultrasonic beam to improve the chances of receiving sufficient reflection from the non-smooth, highly porous, uneven, non-planar, irregular snow surface.

The fourth paper presents the design considerations and performance characteristics of Snow Temperature Profile Sensing System used to measure the temperature gradient and temperature distributions within and outside the snowpack at different depths.

The fifth research paper focuses on describing the design of Snow Temperature Profile Sensing System in details and discusses the theoretical and mathematical model that outline important temperature parameters. Then the paper describes how the system is implemented to record or measure those parameters.

The sixth paper presents the design considerations, constraints and design techniques used to use RTD temperature sensors for snow temperature measurement applications. The paper also presents the performance evaluation and suitability of such sensors.

The seventh paper focuses on design techniques for front-end analog signal conditioning module and the design challenges faced when interfacing analog unit to a data acquisition system.

The eighth paper describes the design of snow air temperature sensing probe and methods to ensure that it measures true air temperature over a snow cover and is not influenced by solar radiations and winds.

The book may be read as an applied text-book in conjunction with standard electronics and instrumentation design textbooks. The book will guide students on how to apply basic principles of instrumentation systems design, integrate concepts of physical sciences and measurement sciences for the field applications.

A NOTE TO READERS

The research papers in this series were authored between 1996 to 2005. As such these papers should be read remembering the time frame in which those were written. Though the basics of design are universal, author has made no claim regarding the contemporariness of the concepts. While the book presents the most fundamental and universally applicable basic principles in electronics, new advances in analog electronics and system design should be considered when using or extending the designs discussed in this book.

Each paper was written for different aspects of the overall system design and has been used in this book as-it-is. A substantial overlap, repetitions of text and redundancies are thus imperative to make each paper read of its own.

ABOUT THE AUTHOR

Raman K Attri is a corporate business researcher, learning strategist, and management consultant. Masters in electronics engineering, he served as an electronics design scientist at a premier research organization. He has served at technical and product development roles at leading international corporations. As an engineer, he specializes in systems engineering of complex equipment, scientific instrumentation sensors and system design. His international professional career spanned over 25 years across a range of disciplines such as scientific research, systems engineering, management consulting, training operations, and learning design. With his technical and training background, he focuses on the competitive strategies to develop the technical workforce with higher-order troubleshooting and problem-solving skills at a much faster rate. He provides strategic consulting to the organizations by accelerating time-to-proficiency of employees through well-researched models. He holds a doctorate in business from Southern Cross University, Australia.

CITATION DETAILS

The collection can be cited as:

Attri, RK 2018, *Research and Design of Snow Hydrology Sensors and Instrumentation*, R.Attri Instrumentation Design Series (Snow Hydrology), 2nd edn., Speed To Proficiency Research: S2Pro©, Singapore.

This series contains these seven papers, which can be individually cited as:

Attri, RK 2018/2005, 'Design of a Reliable Remote Surface Detector Based on Ultrasonic Pulse- Transit Technique to Detect Uneven & Non-Smooth Porous Snow Surfaces,' R.Attri Instrumentation Design Series (Snow Hydrology), Paper No. 1, *Research and Design of Snow Hydrology Sensors and Instrumentation*, 2nd edn., pp. 1-21, Speed To Proficiency Research: S2Pro©, Singapore.

Attri, RK 2018/1999, 'Design strategy of snow depth sensor based on ultrasonic pulse-transit technique for remote measurement of snow cover thickness,' R.Attri Instrumentation Design Series (Snow Hydrology), Paper No. 2, *Research and Design of Snow Hydrology Sensors and Instrumentation*, 2nd edn., pp. 23-41, Speed To Proficiency Research: S2Pro©, Singapore.

Attri, RK 2018/1999, 'Implementation of Linear Array of Ultrasonic Transmitter-Receiver Transducers for Detection of Non-Smooth Porous Surface,' R.Attri Instrumentation Design Series (Snow Hydrology), Paper No. 3, *Research and Design of Snow Hydrology Sensors and Instrumentation*, 2nd edn., pp. 43-66, Speed To Proficiency Research: S2Pro©, Singapore.

Attri, RK 2018/1996, 'Snow Pack Temperature Profile Sensor,' R.Attri Instrumentation Design Series (Snow Hydrology), Paper No. 4, *Research and Design of Snow Hydrology Sensors and Instrumentation*, 2nd edn., pp. 67-76, Speed To Proficiency Research: S2Pro©, Singapore.

Attri, RK 2018/1999, 'Design of an Instrumentation System to Record Distribution Profile of Snow Layer Temperature for Modelling of Snow Avalanche Forecast,' R.Attri Instrumentation Design Series (Snow Hydrology), Paper No. 5, *Research and Design of Snow Hydrology Sensors and Instrumentation*, 2nd edn., pp. 77-94, Speed To Proficiency Research: S2Pro©, Singapore.

Attri, RK 2018/2000, 'Design Approach to Use Platinum RTD Sensor in Snow Temperature Measurements,' R.Attri Instrumentation Design Series (Snow Hydrology), Paper No. 6, *Research and Design of Snow Hydrology Sensors and Instrumentation*, 2nd edn., pp. 95-106, Speed To Proficiency Research: S2Pro©,

Singapore.

Attri, RK 2018/2000, 'Practical Design Considerations for Signal Conditioning Unit Interfaced with Multi-point Snow Temperature Recording System,' R.Attri Instrumentation Design Series (Snow Hydrology), Paper No. 7, *Research and Design of Snow Hydrology Sensors and Instrumentation*, 2nd edn., pp. 107-126, Speed To Proficiency Research: S2Pro©, Singapore.

Attri, RK 2018/2000, 'Design of A True Snow Air Temperature Sensing Probe,' R.Attri Instrumentation Design Series (Snow Hydrology), Paper No. 8, *Research and Design of Snow Hydrology Sensors and Instrumentation*, 2nd edn., pp. 127-137, Speed To Proficiency Research: S2Pro©, Singapore.

PREVIOUS WORK

Author's previous research work on snow hydrology can be cited as:

Shamshi, MA, Attri, RK & Sharma, VP 1996, 'Snow pack temperature sensor,' *Proceedings of National Conference on Sensors and Transducers*, Chandigarh, pp. 180-189, viewed 24 Jan 2018, <https://www.researchgate.net/publication/275276742>.

Attri, RK, Sharma, BK, Shamshi, MA & Sharma VP, 2000, 'Design Approach to use Platinum RTD Sensor in Snow Temperature Measurements', *Journal of Instruments Society of India*, vol. 30, no. 4, pp. 275-283, available at https://www.researchgate.net/publication/275276709.

Attri, RK, Sharma, BK & Shamshi, MA, 2000, 'Practical Design Considerations for Signal Conditioning Unit Interfaced with Multi-point Snow Temperature Recording System', *IETE Technical Review*, vol. 17, no.64, pp. 351-361, https://doi.org/10.1080/02564602.2000.11416928, or download at https://www.researchgate.net/publication/275276698.

DESIGN OF A RELIABLE REMOTE SURFACE DETECTOR BASED ON ULTRASONIC PULSE-TRANSIT TECHNIQUE TO DETECT UNEVEN & NON-SMOOTH POROUS SNOW SURFACES

RAMAN K. ATTRI

EX-SCIENTIST,
CENTRAL SCIENTIFIC INSTRUMENTS ORGANIZATION INDIA

Manuscript originally written Sept 2005

Abstract - **This paper discusses design control techniques used to design the ultrasonic pulse-echo method based remote surface detector to detect objects with porous, uneven, irregular surfaces like snow. For such surfaces, there is an ample possibility of scattering of the reflected beam in multiple directions, absorption of ultrasonic energy within the pores and reflected energy too feeble to detect. Sometimes, the reflected beam may altogether miss the receiver resulting in no surface detection. Scatter and penetration of ultrasonic energy at such irregular porous uneven surfaces poses great difficulties in designing a reliable remote surface detector. In this paper, we have presented a comprehensive set of various design techniques to develop a highly reliable remote surface detector. Experimental investigations have resulted in a highly accurate and reliable snow surface detector device being used in deep Himalayan regions for snow avalanche forecasting equipment.**

Keywords: Ultrasonic, snow surface, range detection, remote sensing, object detection, porous surfaces, piezoelectric sensors

1. INTRODUCTION

Remote Surface Detection (RSD) using Ultrasonic Pulse-Transit (Pulse-echo) method is one of the popular methods for detecting the target surface, distance estimation and 3-D imaging [1]. Most such methods are used for a target having a solid, smooth and possibly a flat surface. For such surfaces, the chances of reflected energy back to the source are high, and the target is detected accurately [6, 7]. There are many applications where the same method of Ultrasonic Pulse-transit has been extended for detection of targets with irregular, non-smooth surfaces, uneven surfaces or porous surfaces [21].

Snow surface detection is one of the very important applications of such ultrasonic RSDs to determine the thickness of the snow layers by finding how much the current surface level snowfall is above the ground level. This requires detection of snow surface detection distance from the sensor [2, 18]. This information is used for critical and man-kind saving hydrological studies such as forecasting of a snow avalanche, river run-off water, glacier sliding, river water levels and related phenomenon in the mountain areas and planes nearby [4]. The fresh snow is porous non-smooth, and the irregular surface where using remote sensing of snow using ultrasonic beams has its own problems [3, 5]. Such surface detection requires a lot many considerations in the design of such ultrasonic RSDs [8]. Such critically irregular and non-smooth porous surface (e.g. Snow surface) causes penetration of the wave of the incident wave into the surface, absorption in the surface and scattering around which results in either missed reflected wave or a very low amplitude highly noise-ridden reflected signal [11, 20, 21].

The author conducted a range of experiments using different design options to develop a highly accurate and reliable snow surface detector device. Because of multiple experiments, a set of design control techniques evolved which proved very effective and likely to apply to other applications involving similar porous surfaces like sand, chemical compound fills, crops fillers etc. In this paper, we will present the tested design techniques found effective for improving sensitivity, range, near-field side-lobe compensation, and temperature-velocity ultrasonic waves.

2. ISSUES WITH PERFORMANCE OF POROUS SURFACE DETECTORS

Experiments revealed several practical issues with conventional remote surface detectors (RSDs) were observed due to characteristic behavior of the snow surface and its porosity; Literature reports several issues on various aspects when of ultrasonic transducers when used for surface detection applications.

The snow typically is highly crystallized but porous surface. At all frequencies of transducer wave, the losses due to absorption are virtually unavoidable [2, 3]. Penetration of ultrasonic waves at high frequency inside the snow surface is one of the biggest problems which require selection of proper ultrasonic generating transducers [11, 20]. The higher frequency causes penetration in the porous snow surface and causes a lot of heating and absorption in the surface [7]. At a lower frequency of the ultrasonic, the penetration and absorption losses are less, but the accuracy of measurement suffers. On the other hand, at a high frequency, the penetration is more, but the range is poor [20]. Therefore, the selection of the right frequency of the ultrasonic transducer is very important for effective measurement.

Temperature dependence of impendence and capacitance of the piezoelectric transducers is also an issue. For the existing application, the transducer is expected to work in -200 C and requires a temperature stabilized transducer [9, 19]. Temperature dependence of sound velocity and change in the directional pattern of the sensor due to fluctuation in temperature is also a big problem [9]. Temperatures may be down to -40°C in the heavily snowbound area.

Another impairing factor comes by a typical ultrasonic beam shape at the transducer. Near-field interference by side grating lobes in the beam triggers the receivers for false near-field object detection [10]. These grating lobes induce a signal in the nearest receiver situated on the side of these lobes; hence it is taken as a reflected signal by the system, thus giving wrong distance reading as if the target surface is very near to the sensor hood [12, 13].

Irregular snow surface shape plays a vital role in the received echo amplitude and shape. It is seen that roughness of over 1/10 of wavelength impairs the coupling markedly and makes ultrasonic beam to become diffused and scatter in all directions [10]. Sometimes the reflected echo is not received. Another issue created by irregular non-smooth surfaces is that more the directive the beam is, more would be the probability of scattering. These directive beams, when reflected from non-even surfaces, may altogether miss the receiving area [12]. The physical design of the receiver with a proper "looking area" receiver is critical to enhancing the probability and hence the reliability of the detector [14]. The biggest issue we faced is what should be the area covered by the beam width on the porous surface to ensure that sufficient coverage of toughs and crest is ensured with a high probability of finding a flat area from where there is a good reflection. Another important impact of surface irregularity seen during the experiments emerges in the form of constructive interference when the incident beam strikes the surface at such an angle that the scattered beams make high amplitude interference pattern at the receiver sensor. This causes detectable echo amplitude even when the surface was beyond the range of the system. This is essentially a false surface somewhere "in-between."

All design improvement apart, the strength of transmitted pulse governs the rest of the design. Higher the voltage at the transmitter, higher is the power consumption which is a concern in such field battery operator devices. Experimentally it has been found that increase in transmitted voltage amplitude does have some improvement of results, but this improvement gets saturated after a certain point, and no amount of increase in transmitted voltage make any effect [1, 7, 11]. Choosing an optimum transmitter voltage level is an important aspect of the design; because this governs the worst-case amplitude received at the receiver and hence defines the loop gain required for reliable operation. This also defines the software and hardware thresholds [11].

Piezoelectric transducers typically act as charging-discharging capacitors with sinusoidal responses and harmonic response to the pulse train. The number of pulses in a single burst (or echo) to trigger sound mechanical waves directly affects the shape and amplitude of received wave envelop. This, in turn, impairs the successful detection of the reflected wave. Using too a smaller number of pulses in the burst makes it difficult for the detector to detect received echo successfully [7].

During experiments, it was seen that the amplitude of the detected received signal was too feeble to detect. An attenuation of around 100db after reflection from such non-smooth surfaces at around 4 meters from the sensor was observed. This reflected echo needs to be amplified to a sufficiently detectable level [22]. Further, the increase in amplification of received echo will only increase the noise beyond the threshold level.

The low amplitude reflected echo is further over-ridden in white noise. Noise amplitude is more than the signal amplitude, so a system with a larger signal-to-noise ratio has to be designed. The inherent residual noise of the piezoelectric transducers itself is a great source of the problem [22]. One important point observed is that despite the accurate filtering of noise, the residual noise is always present at the input of the sensors even when they are not receiving. The reflected echo waveform gets added into it, and if the SNR ratio is not controlled correctly, it makes the detector almost impossible to detect and filter the noise-ridden echo [8].

3. DESIGN TECHNIQUES & RESULTS

We carried out multiple design improvements through experiments. These improvements result in a very reliable and sensitive remote surface detector tested successfully on a porous snow surface. Following are the major design changes.

3.1 Physical and Mechanical Design Improvements

From the experiments, we saw that due to wavelength involved, the accuracy of measurement is high at a frequency such as 50 KHz, but penetration of ultrasonic waves is also high thereby causing absorption losses. Also, the range of measurement is too less at such high ultrasonic frequencies. Using lower frequencies in the range of 20KHZ does have some positive effects of increasing the range, reducing the losses due to scattering and absorption due to the directional pattern but then the accuracy of the measurement suffers. Refer to figure 1 which compares the range, accuracy, and penetration of the ultrasonic beam from piezoelectric sensors at different available frequencies. Penetration is maximum at 50 KHz, and the range is best with 25 KHz [20, 21].

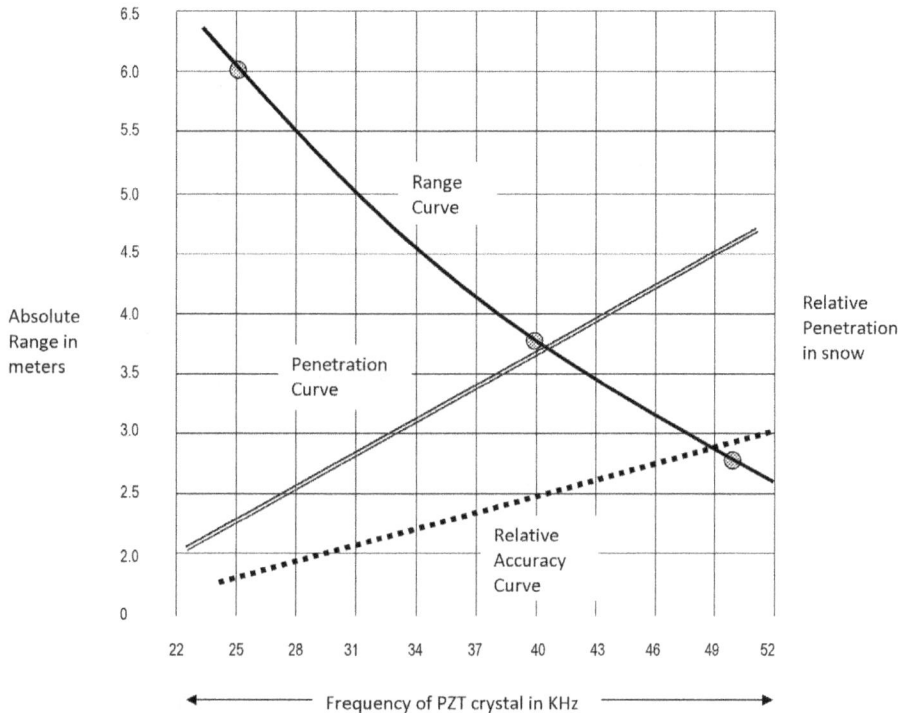

Fig [1]: Graph of values of the range, penetration and relative accuracy of piezoelectric crystals as a function of frequency

Considering the range, accuracy and penetration inside the snow, 40 KHz crystal gives the optimum performance, which has been selected. The availability of lower frequency moisture proof completely sealed piezoelectric crystals working in −30 to +50°C temperature range limited our options. For the present application, we have selected MZT-40E7S-1 from Murata Corp Japan [19]. These are sealed piezoelectric transducer operating at 40 KHz, as shown in figure 2 and are used in dual mode- as receiver and transmitter.

MA40E7R/S

Fig [2]: MA40E7S-1 piezoelectric sensors (Source: Murata Corp Japan) [19]

Beam width plays an important role in such cases [12, 13]. Some observed that sharper beams have more probability of scattering at snow particles and broader beam have better chances of reaching the receiver even during worst-case reflection [26]. The transducer selected for this application has the broader directional response with directivity in sound pressure level of 100db as shown in figure 3 below.

These are dual-use waterproof sensors working at 40 KHz and able to work down to -30°C. Working at a maximum voltage of 100V, the sensor can give 9mm resolution in range detection. With a symmetric directivity of 75°, it generates a broader symmetric beam of 106 dB sound pressure level at 30cm. Although the range of the sensor is specified as 3 meters, however, with design improvements discussed in this paper, we got a range of 4.5 meters combined with a recursive software algorithm for near-field and far-field compensations.

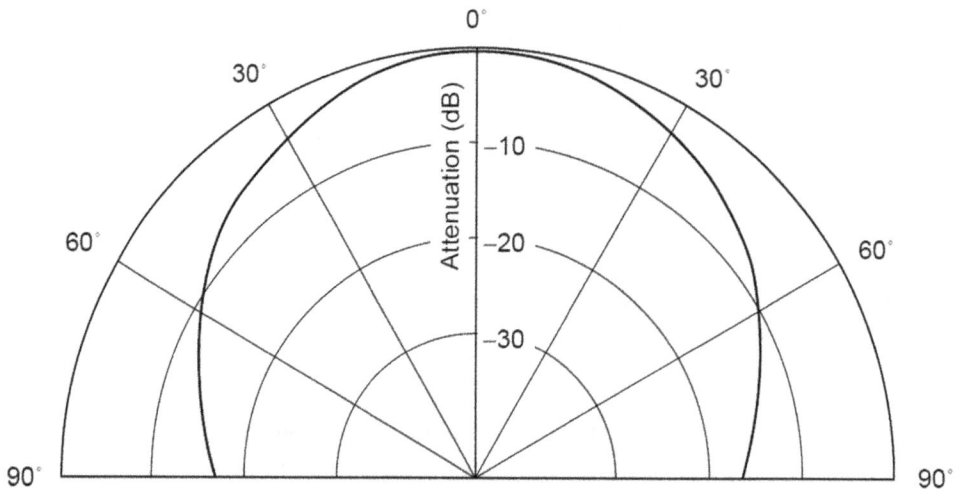

Fig [3]: *Directivity of MA40E7S-1 transmitting piezoelectric transducer in terms of sound pressure level measured in dB at 30cm with 40 KHz frequency. Typical directivity is 75° symmetric in all directions. This results in a broader beam with a broad front zone area of the beam (Source: Curtesy Murata Corp Japan)* [19]

Even if the transducer geometry gave us a broad beam, it was very important to ensure that beam strength density is high at all the point in the front zone of the beam and overall surface coverage area should be broader. To deal with the non-smooth surface of snow and to counteract the scattering effects, the directional pattern of the transmitted ultrasonic beam has been modified using concepts of array theory. According to array theory, the total directional response of an array is simply the sum of directional responses of all the individual transmitting elements [15]. The geometry of these individual elements shapes up the directional pattern, beam width, and side-lobes.

The spacing between the individual transmitting transducers in x-axis and y-axis both affects the directional response. The mounting geometry i.e. whether in a rectangle, square or triangle drastically change the directional pattern [13, 15].

Geometry (i.e., x and y spacing of transducer elements) is found by using basic array theory calculations followed by extensive experiments to fine-tune the right mounting geometry. Based on experiments a line geometry with transmitting array of 3 x 1 (i.e., 3 transducers mounted in a row) having a spacing between them equal to a little less than the half wavelength, was chosen. It is shown in figure 4. This array gave a sufficiently dense and powerful broad beam to avoid any chance to miss the reflected beam on the receiver transducer and to compensate for the losses due to non-smoothness. This design effort increases the possibility of getting the scattered beam even from the non-smooth surface of the snow.

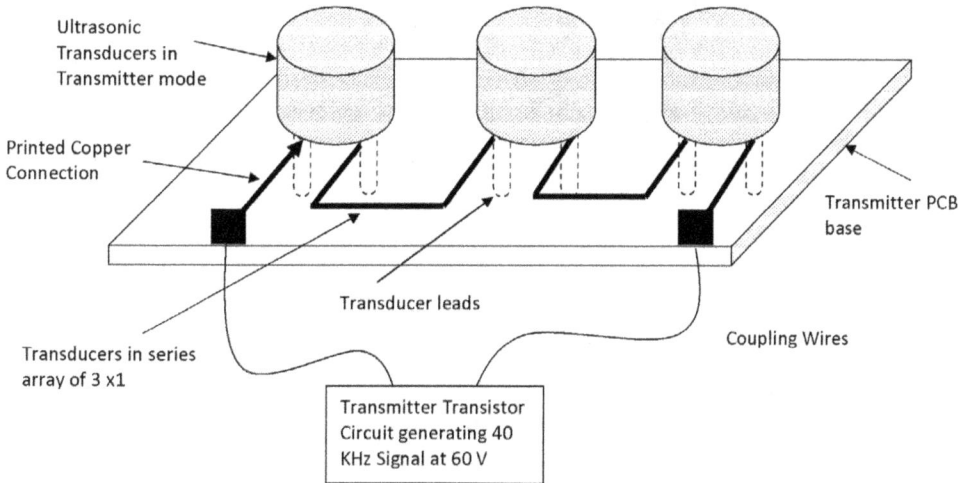

Fig [4]: *3x1 Transmitting transducer connected in a series array to provide superimposed phase-shifted transmitting beam envelope for a high probability of reception*

Losing power distribution over the broader beam width was compensated by increasing the power supply to the transducers from 30V DC to 60V DC [19]. With this configuration, the beam width becomes broader and powerful, so it travels as a broader beam rather than a straight-line thin beam toward the surface. Optimum performance can be achieved by varying the geometrical parameters further and the frequency of transmission if we have an option.

It was found during experiments that to improve the probability of 'worst-case' reflection to be received at the receiver; the receiver face area should be designed correctly regarding transmitter beam width. With experiments, we found that 1:3 ratio of transmitter and receiver works best for optimum but highly reliable performance. We used 1x3 transmitter transducers in an array and 3x3 receiver transducer arrays.

Essentially for one transmitter, there are three corresponding receivers to receive the signal. This essentially increased the probability of reception by 3 times.

The receiver array response was also optimized by using the rectangular geometry consisting of array configuration with separate 3 x 3 receiver array isolated from transmitting array mechanically and electrically, as shown in figure 5. Wider receiver surface area gives the ample opportunity that the reflected beams at some angle will also get captured. The techniques made the system very robust thus worked well with most kinds of rough surfaces and particularly proved suitable on the snow surface.

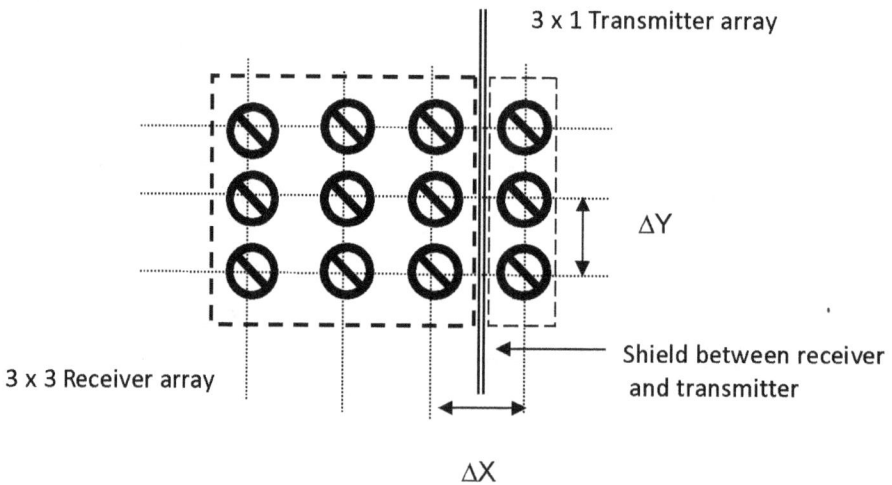

Fig [5]: Array arrangements for transmitter and receiver. Receiver array is a rectangular array of 3x3 and transmitter array is a line array of 3x1.

There are two options of connecting the array: one is putting elements in parallel, and one is putting the array in the series [15]. In our case, we have chosen the latter approach. It has given a big design advantage.

Putting transmitter array in series, one in figure 5 ensured that each of the three transducers are activated with little delay between, so the result is the transmission of three envelopes of bursts. This short delay makes the right superimposition at the rough surface and chances of reflection from any of the envelopes is strengthened. This will be clearer from the figure 6 which shows the three transmitted envelops received at the receiver.

Similarly, the receiver has also been connected in series, as shown in figure 7. This gives a very big advantage that the overall voltage received at the receiver section is the superimposition of the wavefront arriving at each of the receiver transducers. This strengthens the receiver signal and even if only one transducer element has received the wavefront, it acts as the right input signal.

Time shifted Superimposed Envelops

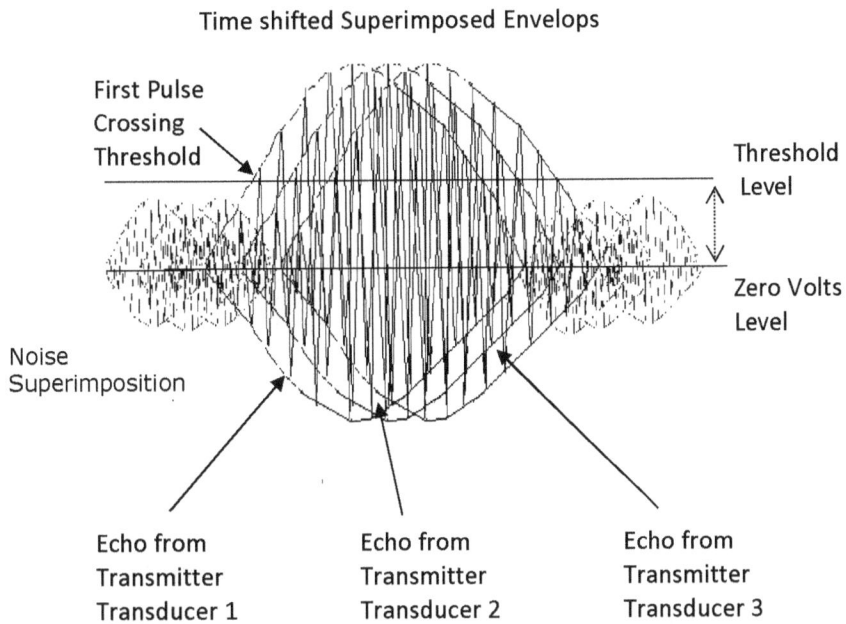

Fig [6]: *Superimposed time-shifted 3 transmitted wavefront envelops. Due to the capacitive effect, the three transmitters get triggered with a short delay in between. This results in a superimposed three envelopes of pulses triggering in sequence*

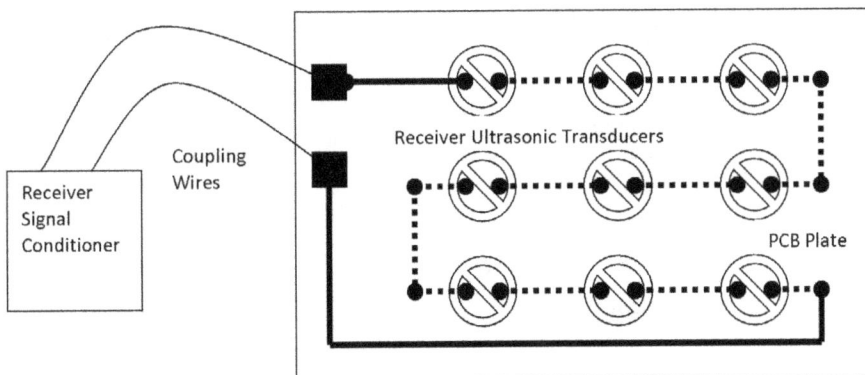

Fig [7]: *3x3 rectangular receiver transducers array connected in a series array. The series connection ensures the summation of the envelops received by each receiver transducer to create a resultant superimposed signal at amplifier inputs*

The receiver section, in this case, will receive the three echoes since there were three transmitted envelopes. This ensures that even if we are transmitting one envelop consisting of N number of pulses, the receiver wavefront will be a superimposed wavefront consisting of 3 x N number of pulses. This increases the system responsiveness and lessens the chances of missing the reflection.

3.2   *Electronics Circuit Design Improvements*

In the pulse-transit method, only a short burst of pulses is sent toward the target. Using a single pulse is not effective in a situation as there is a good probability of a single pulse distortion. Even when multiple pulses were used the inherent damping factor of the transducer at the transmitter side and receiver side made these pulses to exhibit modulated envelope with a peak at the center of the envelope. The pulses at the extremes of this have low amplitude [1].

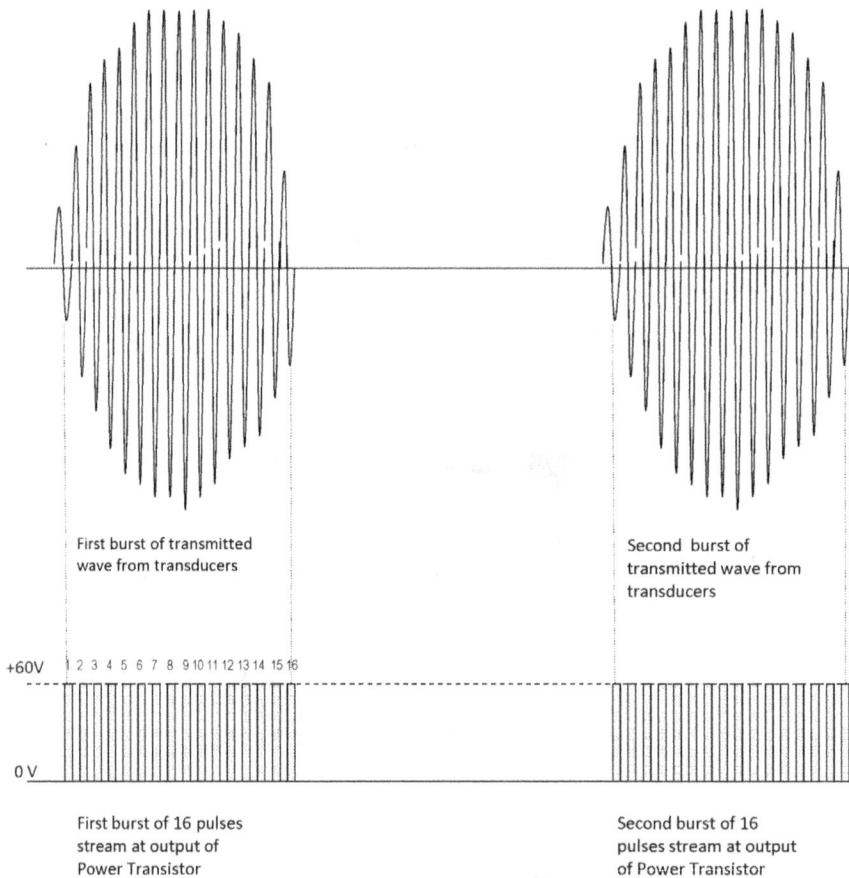

First burst of transmitted wave from transducers

Second burst of transmitted wave from transducers

+60V   1 2 3 4 5 6 7 8 9 10 11 12 13 14 15 16

0 V

First burst of 16 pulses stream at output of Power Transistor

Second burst of 16 pulses stream at output of Power Transistor

Fig [8]: *A burst of 16 ideal square pulses of 60V triggered by output amplifier produces modulated mechanical waves at the transmitter transducer. The received signal exhibits similar characteristics.*

For successful detection of the received echo, it is very important that the amplitude of at least one of the received pulses in the burst should be higher than the set threshold. Refer to figure 8, which shows what is being transmitted.

Experiments showed that 16 numbers of pulses were optimum for detection of the porous snow surface. We used an electronic circuit to control transmission of 16 pulses. A single pulse triggers the transmission section. The transmitter operated on 40 KHz frequency generated by 40 KHz oscillator. An R-S flip-flop in conjunction with the 16-pulse counter was used to transmit only 16 pulses to the base of the Transmitter Amplifier as shown in figure 9. The duty cycle of 25% was chosen to give a tolerance for the transition time of transistors and rise time of ultrasonic transducers.

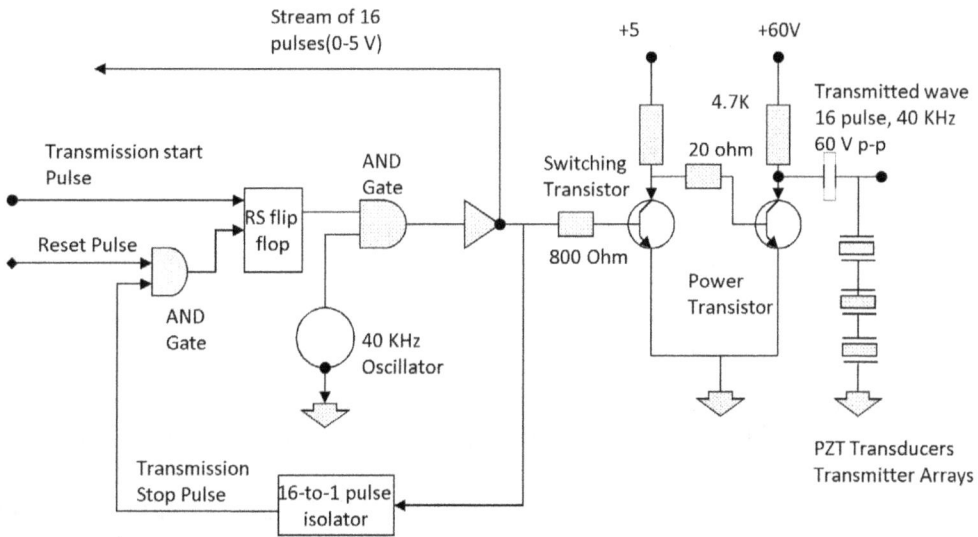

Fig [9]: Transmitter Section is a Class-C amplifier operating at 60V, driven by a switching transistor. An oscillator and RS flip-flop combination logic ensure the passage of 16 pulses to the power amplifier and to a series array of transducers.

The primary response of the remote surface detector depends upon the amplitude of the transmitted signal. Greater the amplitude better shall be the reflected echo. Normally piezoelectric ultrasonic transducers are operated at high voltages as high as 100 V. Experimentally it was observed that an increase in transmitted amplitude has a very minimal effect on the response of the system. With an increase in 10% of transmitter amplitude voltage, only 1% improvement in the response is achieved [11]. So, the transmitter voltage could not be raised beyond a threshold, if raised at all, the improvement of response was minimal at the cost of heavy current consumption. 60V was chosen as optimal transmitter voltage.

The output of the receiver piezoelectric sensor is as minute as 1 mV in response to the 60 V transmitted signal. This reflected echo is shadowed so severely by the noise of higher amplitude that it was impossible to detect it without further signals conditioning. The reflected echo is applied to the input amplifier stage which server the dual purpose of impedance matching and initial gain. Refer to figure 10 showing the receiver circuit diagram.

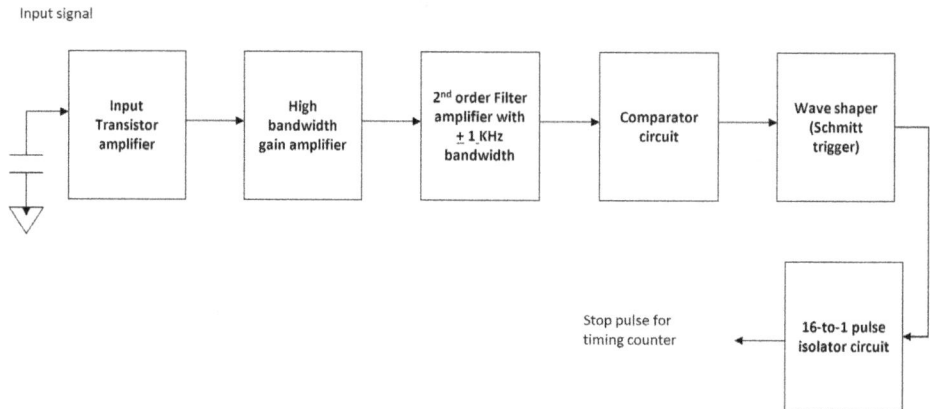

Fig [10]: *Block diagram of the receiver wave-shaping circuit. The circuit amplifies the weak received signal, compare it with a nominal threshold to eliminate the white noise and shape the resultant signal into square wave pulses*

Inputs stage gain was designed at 200. The amplified signal was still in mV range in figure 11a and was highly noise-ridden spanning over the entire spectrum. A high gain amplifier with an optimum gain of 50 was used for further amplification of the signal. The useful signal bandwidth is approximately 50 kHz, which implies unity gain-bandwidth of approximately 2500 kHz. A video amplifier with UGB of 120 MHz has been used to amplify all the frequencies uniformly. The output signal of the order of 1 V was received but still shadowed by the noise of higher amplitude. As shown in figure 11b the reflected echo is still not reproduced faithfully.

Experiments show that the high amplitude noise is well above or well below the center frequency of 40 KHz. A precision filter amplifier was designed as a narrow band-pass filter with sharp cut-offs at 38 kHz and 42 kHz with a center frequency of 40 KHz. The response of this filter stage is shown in the figure 11c. The resultant echo of very clear shape and amplitude was received.

The number of pulses received in this echo is three times the number of transmitted pulses (because of three transmitters). Actually the received echo consists of three envelopes shifted slightly in phase and each envelope having 16 pulses. The increased number of pulses gives better chances of triggering of the next circuit and hence the measurement and hence better sensitivity.

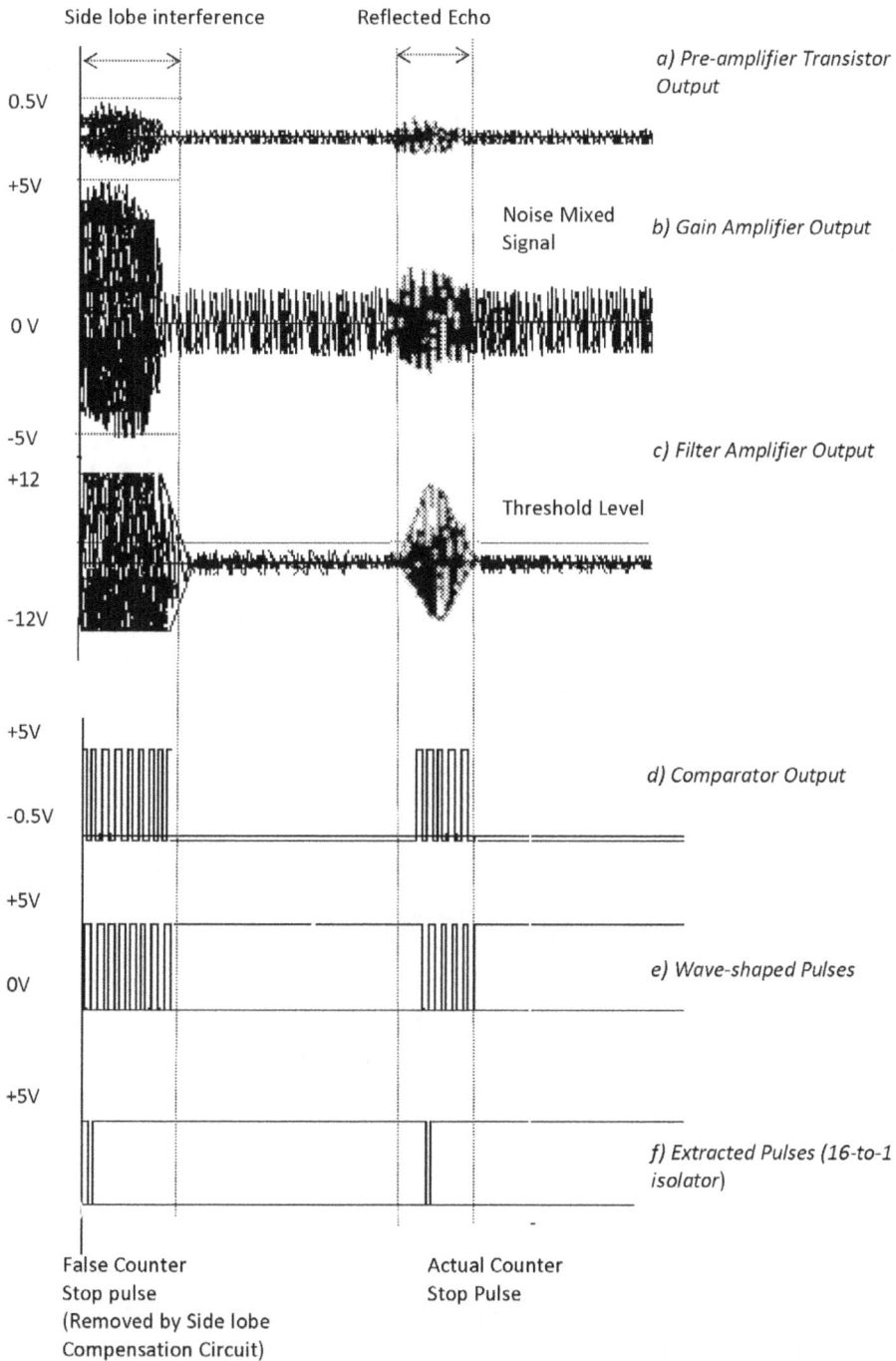

Fig [11]: a) Input signal at pre-amplifier transistors with very low amplitude ridden in the mechanical noise b) Output signal from Gain Amplifier which amplifies the weak signal and amplifies the background noise. The prominent signal level corresponds to the reflected echo. First saturated signal corresponds to either side-lobes or false trigger due to broader beam c)

Output signal at Filter Amplifier output which cleans the signal and removes the noise. The reflected echo signal is enhanced for digitization. d) Pulse stream at comparator which removes the background residual noise and digitizes the inputs signal peaks above the threshold voltage e) Pulse stream at Wave Shaper Output removes the negative side of the pulses and produces clean rectangular pulses from 0-5V f) Output of 16-to-1 pulse isolator circuit which produces one pulse against a stream of 16 pulses. This single pulse is used to stop the timing counter for measuring the transit time

One important point observed is that despite the accurate filtering of noise, the residual white noise is always present at the input of the sensors even when they are not receiving. This residual noise gets amplified and appears at the output of the final filter amplifier and is present all the time. The reflected echo waveform gets added to it. Generally, cross convolution techniques are employed to remove such residual noise. In the present case, an alternative method of comparison with some fixed threshold has been used [23]. The average value of residual noise is checked out.

A comparator circuit as shown in figure 11c is used to compare the filtered output to this threshold voltage. If the reflected echo above this threshold, comparator gives the output in the form of pulses crossing the threshold mark. The comparator also wave-shape the bipolar sine signals into unipolar square pulses. One important precaution in this design is setting gain and the threshold value of the circuit. A tradeoff point has been observed beyond which accuracy may get impaired if we further lower the threshold and increase the amplification to increase the system sensitivity.

The result of using the comparator is that by setting the threshold correctly, we can ensure to receive the trigger from the genuine echo signal. The effect of threshold setting can be seen in figure 11d, which helps to detect only the meaningful signal above the threshold and hence discards the residual noise.

The pulses received are conditioned and wave-shaped to drive the digital circuit, refer to figure 11e. The first pulse out of an incoming pulse stream pattern as shown in figure 11f is extracted as only this pulse corresponds to the arrival of the echo itself. The first pulse extracted from transmitted stream starts a timing counter, and this first pulse of an incoming stream stops this counter. The counts in the counter are a direct indication of flight time (time taken by the echo to travel to the target and return back to the sensor). This time is used in the computation of the distance of the target surface from the sensor.

While testing the system, it was observed that the two streams of the received pulses are obtained, first one with a large amplitude. The first stream comes immediately after the transmission and another stream arrives late. The first one causes a false triggering of the receiver and causes the time counter to stop, because of which system presume there is an object in the close proximity. Investigations showed that this is because of one of the two factors:

First, the broader beam dispersion of the transmitting beam (typically 75°) may trigger the receiver transducers spaced marginally. Even when a shield is applied between the transmitter and receiver pairs, there are good chances of false triggering from the beam edges touching the receiver reception area. The effect is shown in figure 12.

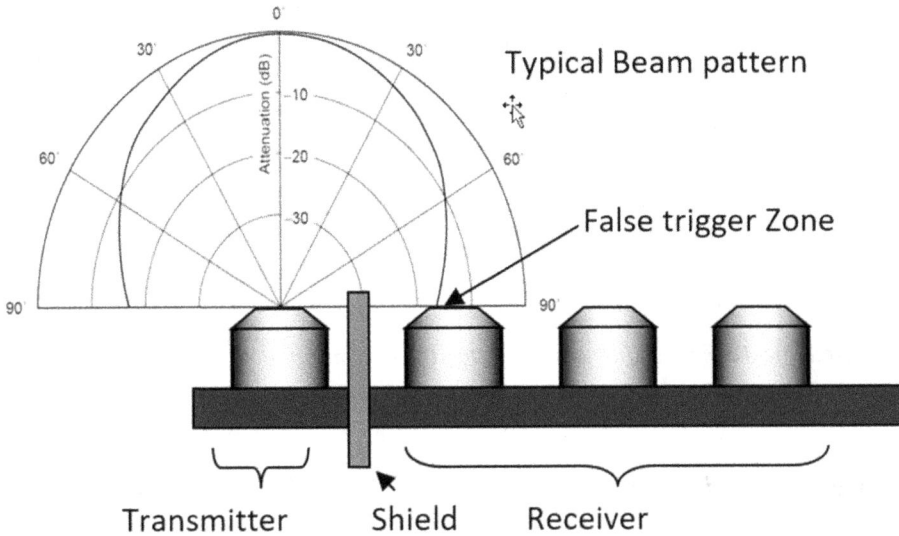

Fig [12]: *Near-field interference due to broader transmitter beam or due to side grating lobes.*

Second, this false trigger could be caused by the side grating lobes of the beam. Although as per manufacturer data, the directional pattern has no side-lobes, but the effect of temperature on changing the directional pattern at -30°C is not known. Irrespective of the source of this interference (whether due to broader beam or due to side-lobe), the counter-measure is same.

As a counter-measure to stop false detection, we developed a circuit and software design to block the first stream of pulses. The figure 13 shows the circuit for blocking the first stream of pulses till the total worst-case duration of the first stream so that the sensor responds to actual echo only. The voltage level at the input of OR gate is made high till the expected time of first steam. The expected time has been around 2300 microseconds. A safe margin is given and pulses till the time 2560 microsecond are blocked. After that, any stop pulse arriving is passed through and the stop the counter. In software, any reading below counts corresponding to this time are rejected, and measurement is termed false, and the cycle is again repeated.

The downside to this implementation is that the sensor can not effectively detect any surface within 50cm. This is called a blind range. However, this was acceptable in our case, since snow surface is never likely to touch the sensor hood.

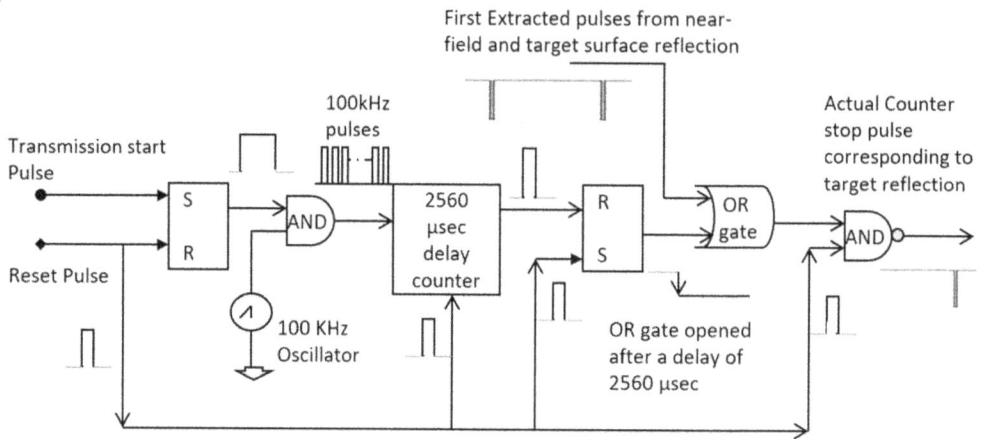

Fig [13]: *Near-field compensation circuit to stop the system counter to avoid false triggering of receiver circuit due to broad beam pattern from the transmitter and/or side grating lobes.*

3.3   Compensation and Improvements Through a Software Algorithm

Due to irregularities of the surface, the reflected beam may miss the receiver, and in the system, the counter may keep waiting for the reflected echo. To bring the system out of wait mode, software checks for the maximum expected counts. If current counts cross the maximum permissible counts, the system takes the situation as a missed beam and repeat the measurement cycle until it gets the counts in the permissible accurate range. After repeating the cycle for 3-4 times, if a valid reading is not available, the system indicates the message on the LCD that no target surface found within range. This warning message allows the operator to take corrective actions.

There may be a kind of constructive interference patterned formed at receiver due to superimposed scattered beams. The reason for this is not fully known, but these results in some inconsistent measurements. We postulated that it probably is due to the broadening of the beam and hence multiple reflections from multiple sides. To create the repeatability in the readings and to remove such inconsistent measurements, through software algorithm the same measurement is repeated thrice in succession and a comparison is made if all the three readings are in close proximity. If they are in close proximity, the average of the distance measurement readings is taken, and hence repeatable results are achieved. If the readings are not in close proximity, then software logic discards this as false triggering and repeats the cycle. This far-field compensation enhances the system performance [8].

The velocity compensation has to be provided in the design of the system. To counter this dependence, an integrated solid-state temperature sensor has been used along with its linearized circuit, which reads the temperature with the resolution of 0.01° Centigrade [16, 17]. 8-bit analog to digital converter (ADC) is used for digital conversion of temperature. Extensive Software computations have been carried out to compensate the effects of the temperature over the velocity of the sound.

The general block diagram of a remote surface detector (RSD) is given in figure 14. It is basically a microprocessor-controlled unit with transmitter and receiver section. One section handles side-lobe compensation. The temperature compensation is provided as well to compensate for velocity variations with temperature. The computations are done in software code to make necessary adjustments.

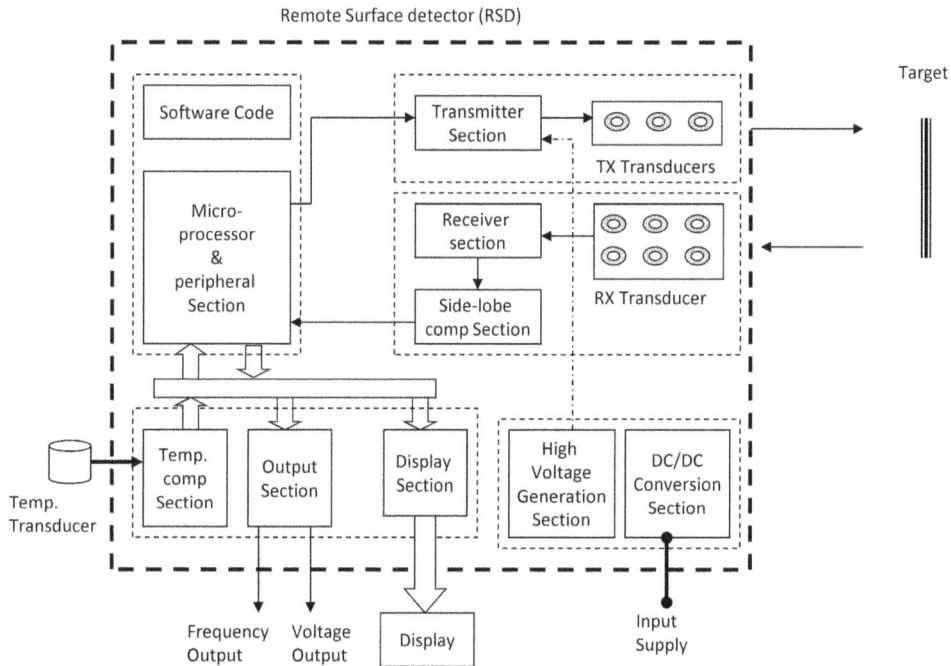

Fig [14]: *General block diagram of remote surface detector*

4. EXPERIMENTAL SETUP OF REMOTE SURFACE DETECTOR (RSD)

A test setup was erected as shown in figure 15 inside the cold chamber controlled at a specific temperature near 0°C. The fresh snow cover was collected and dispensed using a snow dispenser over a non-movable platform of 25 square meters. The test platform of dispensed/ scattered fresh snow inside the cold chamber gave close to the realistic test setup. The care was taken not to make the fresh snow flattened on the platform. The platform was basically a cuboids' container with a height of 10cm. The dispenser filled the container until the top edge with substantial pores, irregularities, and unevenness.

For measurement purposes, the height of snow cover top surface was taken as the height of the platform edge, i.e. 10 cm.

The RSD was mounted on a rigid mast of 5-meter height with the sensor facing downwards toward the platform filled with fresh snow. The typical mounting setup is shown in figure 15. The RSD mast movement was controlled by a hydraulic system to lower its height vertically downwards toward the platform in the increments of 1 cm. The physical distance of the top surface of the snow cover on the platform from the sensor was calibrated initially. The reading on a distance of snow cover from the transducers as reported by RSD was stored along with the physical distance of the RSD from the snow cover.

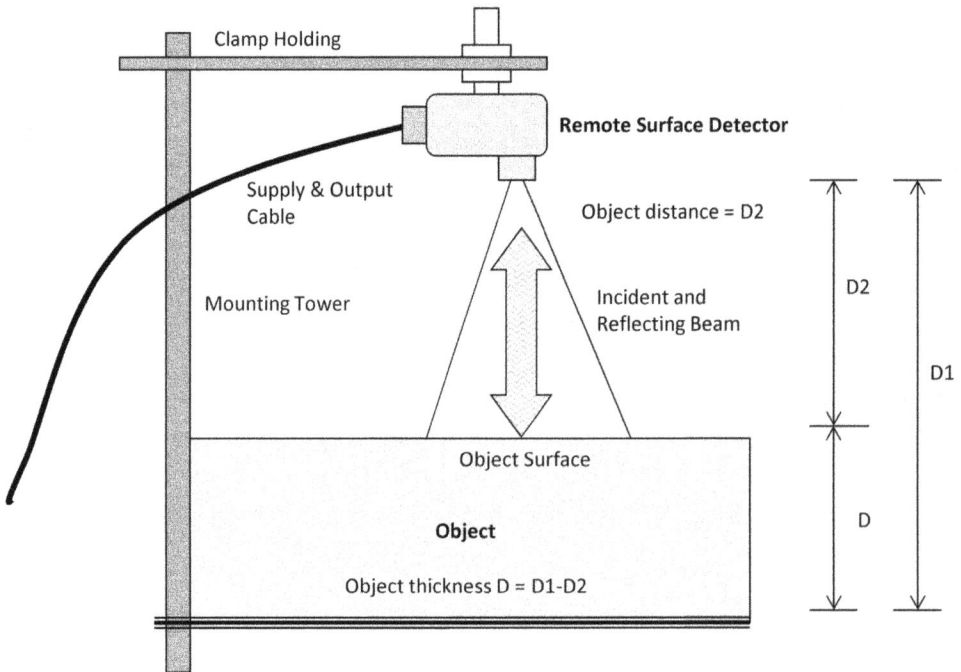

Fig [15]: *Remote surface detector mounting arrangement*

5. DISCUSSION ON SYSTEM PERFORMANCE

The system has been designed to work not only for sampling-based measurements but also for a continuous round the clock functioning. The long-term stability of the readings is checked by recording the distance shown by this system continuously for 1 hour and averaging those out to a single number was taken as the distance reported by RSD. The error between RSD reading and actual reading is plotted along Y-axis in figure 16. Actual distance as recorded physically is plotted against Y-axis.

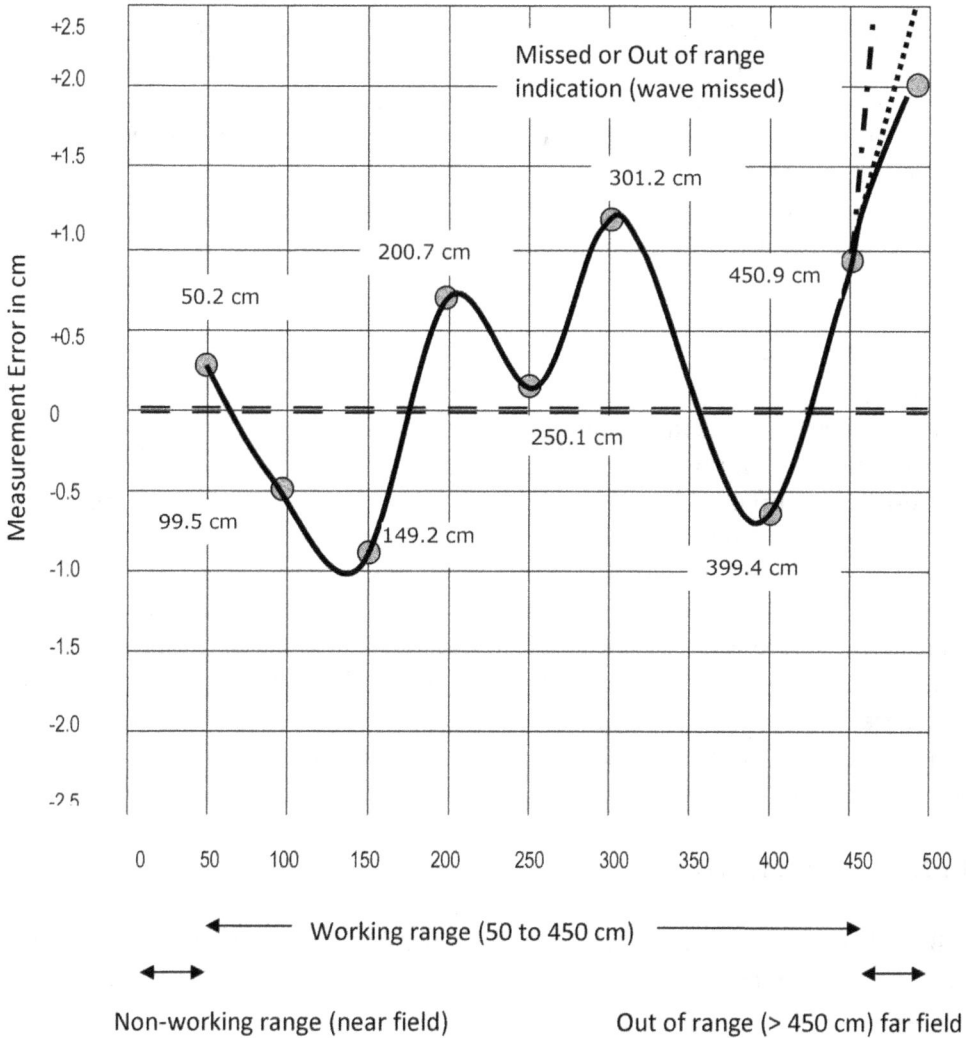

Fig [16]: *System performance diagram showing error in cm against the zero-error axis at different points within the range. The system records distances between 50cm and 450 cm accurately. As it moves away from 450cm, the scattering effects come into the picture which may make the reflected beam to miss the receiver altogether. The accuracy is impaired in the far-field. Detection of the surface is inhibited up to 50cm in the near-field due to side-lobe compensation (the captions on the points indicate the measured absolute distances in cm).*

The accuracy of ± 1 cm with the resolution of 1 cm is achieved within the range of 4.5 meters. On smooth flat surfaces, system measure up to 5 meters with ± 2 cm accuracy. System performance over all kind of porous and irregular surfaces has been checked out and has been found very reliable, repeatable and accurate within ± 1 cm up to 4.5 meters of range. Due to reinforced array technique and software repeated measurements, optimum spacing, and broadening of the beam; the reliable measurement has been achieved up to 4.5 meter. The wave is sometimes missed and not received back as an echo if the sensor is away from the target surface over 450 cm (far-field). System software catches such events and repeats the measurements and average out the false detections. On the other hand, the near-field error comes in the picture due to side-lobe compensation and the system detects no surface when it is near 50 cm to the sensor face. The side-lobe compensation circuit sacrifice near-field 50 cm measurement range, which practically is never used. The snow surface hardly reaches within 50 cm proximity of the sensor.

6. CONCLUSION

The system has specifically been used for snow depth measurement which is extremely important for modeling of snow avalanche forecast and other related studies. The fresh snow surface being the most porous surface and non-smooth, the design approach for a snow surface would apply to all the other materials. Because of this, the system can be used for industrial, commercial, defense applications. The experimental results have been very favorable. A considerable amount of reflected signal is received through arrayed receivers, which is detected with the help of the electronics design. The sensitivity, accuracy, long-term stability and range of the system have been enhanced using these design techniques. The surface distance over 5 meters can be fairly detected despite the uneven, rough, porous non-smooth graining of the surface which normally does give problems of scattering and absorption of energy into the material.

ACKNOWLEDGMENTS

Dr. B.K. Sharma, Head of Dept, Geo-Scientific Instruments Division, CSIO, Chandigarh Swaranjit Singh, Senior Technical Officer, Geo-Scientific Instruments Division, CSIO, Chandigarh

REFERENCES

[1] G.L Gooberman., Pulse Techniques, Ultrasonic Techniques in Biology and medicine, Illiffe books Ltd, London, 1967.
[2] S. Kumar et al, Snow Depth Senor, Proceeding of national symposium on Sensors and Transducers, 1996.
[3] M. Mellor, Engineering properties of snow, Journal of Glaciology, volume 19, pg 15-66, 1977.
[4] T.J. Yamazaki Kondo, T. Sakuraoka and T. Nakamura, A one dimensional model of evolution of snow cover characteristics, Journal of Glaciology, Vol. 18, pg 22-26,1993.
[5] K. D. Hall, Remote sensing of Ice and Snow, London, Chapman and Hall Publications, 1985.
[6] M.C. Combs; Jr. Goodwin, H. Perry, Adjustable ultrasonic level measurement device, United States Patent 4221004, Aug 1978.
[7] M. Krause, et. al. Comparison of Pulse-Echo-Methods for Testing Concrete, E-Journal of Non-destructive testing, Vol.1 No.10, October 1996.

[8] A. Hämäläinen and D. MacIsaac, Using Ultrasonic Sonar Rangers: Some Practical Problems and How To Overcome Them, Phys. Teach. Vol 40, pp 39, 2002.

[9] D. Zhang D, G.M. Crean, Non-linear acoustic properties of high purity quartz as function of temperature, Proceedings of Development in Acoustic and Ultrasonic, IOP Physical Acoustic Group, Leeds, pp 219-224, Sept 1991.

[10] D.S. Balantine et al, Acoustic Wave Sensors-theory, design and physico-chemical applications, Academic Press, 1997.

[11] D. Ensminger, Ultrasonic – the low and high intensity applications, Marcel Decker Inc, New York, 1973.

[12] I Busch, E. Huxson, Ultrasonic, Encyclopedia of applied Physics, VCH Publishers, vol 1, pp 63-88, 1991.

[13] Ultrasonic, Encyclopedia of Science and Technology, Vol 12, pp 662-664, Mcgraw Hill, 1982.

[14] E.P. Papadakis, Physical acoustic principles and methods (W.P. Mason, ed.), vol 4 B, Academic, NY, 1968.

[15] J.E. Hudson, Adaptive Array Principles, Peter Peregrinus , London, 1981.

[16] P. Klonowski, AN-273: Use of the AD590 Temperature Transducer in a Remote Sensing Application, Analog Devices Application Notes, www.analog.com.

[17] M. P. Timko, A two-terminal IC temperature transducer, IEEE Journal of Solid-State Circuits, vol. SC-11, 1976, pp. 784-788.

[18] SR-50 Sonic Distance Sensor, Campbell Inc, Canada, http://www.campbellsci.com/documents/lit/b_sr50.pdf

[19] Piezoelectric Ceramic Sensor (Piezoliote), Cat-P19-E9, Murata Manufacturing Co. Ltd, Japan, http://www.murata.com/catalog/p19e.pdf.

[20] M.G. Silk, Ultrasonic transducers for non-destructive testing, Adam hilger Ltd, Bristol, 1984.

[21] L. Krautkramer and H. Krautkramer, Ultrasonic testing of materials, Berlin, Springer, 1969.

[22] S Raman et al, Processing of Ultrasonic signal for transducer characterization and for improving signal to noise ratio, Indian Journal of Technology, Vol 31, Nov 1993, pp 774-776.

[23] J.L. Lawson and G.E. Uhlenbeck, Threshold Signals, Mcgraw Hill, 1948, pp 211.

[24] R.K. Attri, 'Design strategy of snow depth sensor based on ultrasonic pulse-transit technique for remote measurement of snow cover thickness,' R.Attri Instrumentation Design Series (Snow Hydrology), Paper No. 2, *Research and Design of Snow Hydrology Sensors and Instrumentation*, 2nd edn., pp. 25-41, Speed To Proficiency Research: S2Pro©, Singapore, 2018/2000.

[25] R.K. Attri, 'Implementation of Linear Array of Ultrasonic Transmitter-Receiver Transducers for Detection of Non-Smooth Porous Surface,' R.Attri Instrumentation Design Series (Snow Hydrology), Paper No. 3, *Research and Design of Snow Hydrology Sensors and Instrumentation*, 2nd edn., pp. 43-57, Speed To Proficiency Research: S2Pro©, Singapore, 2018/1999.

Paper No.2

DESIGN STRATEGY OF SNOW DEPTH SENSOR BASED ON ULTRASONIC PULSE-TRANSIT TECHNIQUE FOR REMOTE MEASUREMENT OF SNOW COVER THICKNESS

RAMAN K. ATTRI

EX-SCIENTIST,
CENTRAL SCIENTIFIC INSTRUMENTS ORGANIZATION INDIA

Manuscript originally written Aug 1999

Abstract - Snow cover thickness is one of the important parameters used in forecasting models for snow-melt, snow run-off water, snow avalanche release, and other snow hydrological changes. Ultrasonic pulse-transit method is being used for such applications universally. Reflected echo coming after reflection from the highly irregular and non-smooth porous surface, is very low amplitude noise-ridden signal. This received echo signal has to be conditioned to remove signal and processed to increase its amplitude to make it sufficiently detectable and to increase the probability of receiving back the reflected echo. A special design of snow depth sensor based on Ultrasonic Pulse-transit Method to improve the sensitivity, reduction of losses has been worked out. This paper discusses the signal processing technique and electronics design approach for the development of this snow depth sensor.

1. INTRODUCTION

The importance of snow hydrological studies and measurements have been recognized globally, and an integrated approach to the study of all the interdependent parameters is underway, and related sensors are being developed [1]. The snow cover thickness is one of the most important hydrological parameters, is used in the forecasting model for snow avalanche release, river run-off water, glacier sliding and related phenomenon in the mountain areas and planes nearby [2, 22]. In the deep snowbound areas, the snows thickness changes continuously because the radiation and climatic changes. The snow cover thickness is an indication of snowbound water.

A snow depth sensor system has been designed [3] to measure the snow depth in highly snowbound areas. This system records the snow depth after a suitable interval. This data is integrated with the automatic weather station to study the variations in the snow cover thickness, net melting, and run-off water.

1.1   Pulse-transit Method Employed in Snow Depth Sensor System

The thickness of the snow cover is found with the help of pulse-transit method [4]. A short burst of ultrasonic pulses is transmitted by the piezoelectric transducer transmitter, which is mounted on a with the sensors facing vertically downward toward the snow. Refer to figure 1.

The transmitted beam strikes the surface of the snow. Some energy gets reflected back and is received by the receiver. The time of travel between the transmission and the reception of the pulses is computed which gives the distance of snow cover from the sensor as per distance-velocity equation. The mounting height (a distance of the sensor from the ground) is already known. By subtracting the measured distance from the installation height, snow depth is obtained.

Snow has a critically irregular and non-smooth porous surface, which cause penetration of a wave of the incident wave into the surface, scattering around and absorption in the surface results in very low amplitude highly noise-ridden reflected signal [21]. This received echo signal has to be conditioned to remove signal and processed to increase its amplitude to make it sufficiently detectable. Sometimes the reflected echo is not received back at all and design has to be carried out to increase the probability of receiving back the reflected echo. A special design has been worked out to compensate for the losses due to snow.

This paper discusses the design of snow depth sensor to improve the signal processing and sensitivity and a range of the system. A further design for near-field compensation and compensation for a variation of sound velocity with temperature is also highlighted.

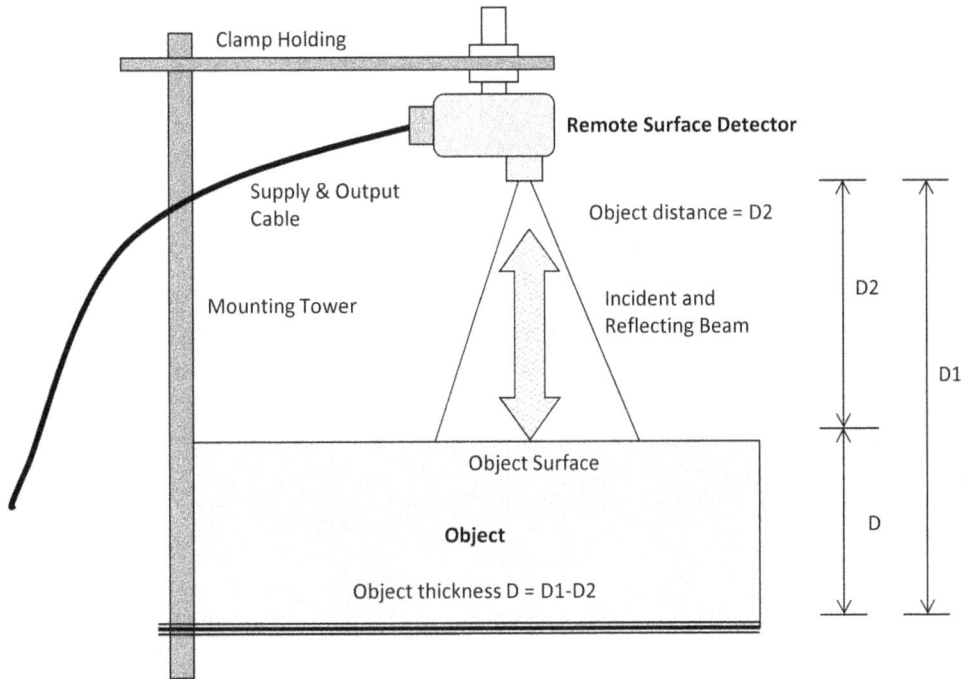

Fig [1]: Snow depth sensor mounting arrangement

2. DESIGN CONSTRAINTS

This simple pulse-transit method of finding the thickness is not as simple when applied to the snow. There are many factors like losses in the snow, the temperature at the place of the measurement and irregularities of snow surface, which govern the range and reliability of the snow depth sensor. These have to be considered while designing the system.

Extreme Environmental constraints: One of the biggest problems in the system design is the temperature and environment specification. Temperature encountered at the place of installation is of the order of -40oC and relative humidity of 100% in a heavily snowbound area. Dependence of ultrasonic velocity on temperature and change in a directional pattern of the sensor due to fluctuation in temperature is also a big problem.

Low Amplitude of reflected echo: The overall performance and reliability of the snow depth sensor based on this method depend upon the ability of the system to detect the reflected echo of the transmitted signal. Further, the system performance can be increased only if one can enhance the reflected echo to a sufficiently detectable level.

Irregularity and non-smoothness of snow surface: In this application, the shape and the roughness of the surface are of decisive importance. These factors often limit the sensitivity of snow depth sensor. A roughness of over 1/10 of wavelength impairs the coupling markedly. The rough surfaces make the ultrasonic beam to become diffused and scatter in all directions.

The porosity of snow cover: The porosity of the snow causes a large amount of ultrasonic energy to get absorbed in the snow, so only a small part of it gets reflected back. The received echo is very weak which need a considerable amount of amplification. The worst of all is that the ultrasonic beam is striking the highly non-smooth surface, which causes the scatter, and the missing of a reflected beam. As a result, the system has to be very sensitive to detect the weak reflected echo. The reflected echo is around 0.1 mV in response to the transmitted wave of 100V.

Noise-ridden signal: The low amplitude reflected echo is further over-ridden in white noise. As the ultrasonic wave is a mechanical wave, so signal noise is very much mechanical in nature spread over the entire spectrum. Noise amplitude is more than the signal amplitude, so the system with a larger signal-to-noise ratio has to be designed. The inherent residual noise of the piezoelectric transducers itself is a great source of the problem.

The worst of all is that sometimes the reflected echo is not received at all. The reflected beam reflects at such an angle that it completely misses the receiver sensors and system remains in wait mode.

Experimentally it has been found that an increase in transmitted voltage amplitude does have some improvement of results, but this improvement gets saturated after a certain point, and no amount of increase in transmitted voltage make any effect. Further, the increase in amplification of received echo will only increase the noise beyond the threshold level. Using lower frequency does have some positive effects of increasing the range and reducing the losses due to scattering and absorption, but there is a limit on lower frequency ultrasonic transducers available.

So, a hardware-software co-design approach has been worked out to modify [1] the directional pattern of the transmitting beam in such a way to reduce the scattering effect and attenuation of the reflected beam. This design further increases the sensitivity of the system by signal processing to extract the useful reflected echo and detection of missed reflected echo.

Inherent errors associated with an ultrasonic beam: Along with design constraints, some errors are also encountered in the system. It pertains to the near-field interference and grating lobes caused by this interference [5]. The reflected echo is received because of these grating lobes instead of the actual reflection, thereby causing the effect of side-looking, implying as if the target is lying in the immediate proximity of the transducers [6]. This near-field compensation has to be provided. The other error is caused by variation of ultrasonic velocity with the temperature. This velocity compensation has to be provided in the design of the system.

3. OVERVIEW SNOW DEPTH SENSOR SYSTEM ARCHITECTURE

The block diagram of the system is given in figure 2. The snow depth sensor has three sections namely I) Transmitter Section ii) Receiver Section iii) Control Section

The system is auto power on triggered and hence needs no switches and buttons to control the system. This will give the measurements as soon as the power is supplied to the system and every 5-Sec after the power on. The power is usually provided by the data acquisition system depending upon its requirement of measurement and its sampling interval. As soon as the reading is taken, power is switched off. The circuit operates under the total control of software. All the pulses are generated by software in conjunction with the hardware circuitry for triggering of the measurements through peripheral devices. It contains the required circuitry needed for ultrasonic transmission and signal conditioning of received ultrasonic signal. This includes the input impedance matching circuit, high gain amplifiers, and narrow band-pass filters. Further, the received pulses are shaped and converted into a suitable format. The near-field compensation circuit is also incorporated here.

Temperature transducer AD590 has been used in this board to sense the temperature and provide corresponding voltage, and ADC is used for converting this voltage into a digital word and compensates the sound velocity. This board also contains the DAC to convert computed results into a real-world analog and frequency signals for interfacing with any kind of data-loggers. 12-bit ADC and 12-bit DAC is incorporated to get sufficient accuracy of measurements.

The dual output is taken from the system. One is the analog voltage in the range of 0-1 V, and other is the frequency in the range of 0-10 kHz in proportion to the distance in the range 0- 10 meters. This output is interfaced to data collection platform or a data acquisition system for further processing [20].

Provision for testing modules is also provided and to take manual measurements. Raw data is further computed using extensive software techniques to convert it into the usable data output.

The power supply card of this sensor also takes care of generating high voltage needed for operation of the piezoelectric transducers, also for providing a conventional power supply for the system.

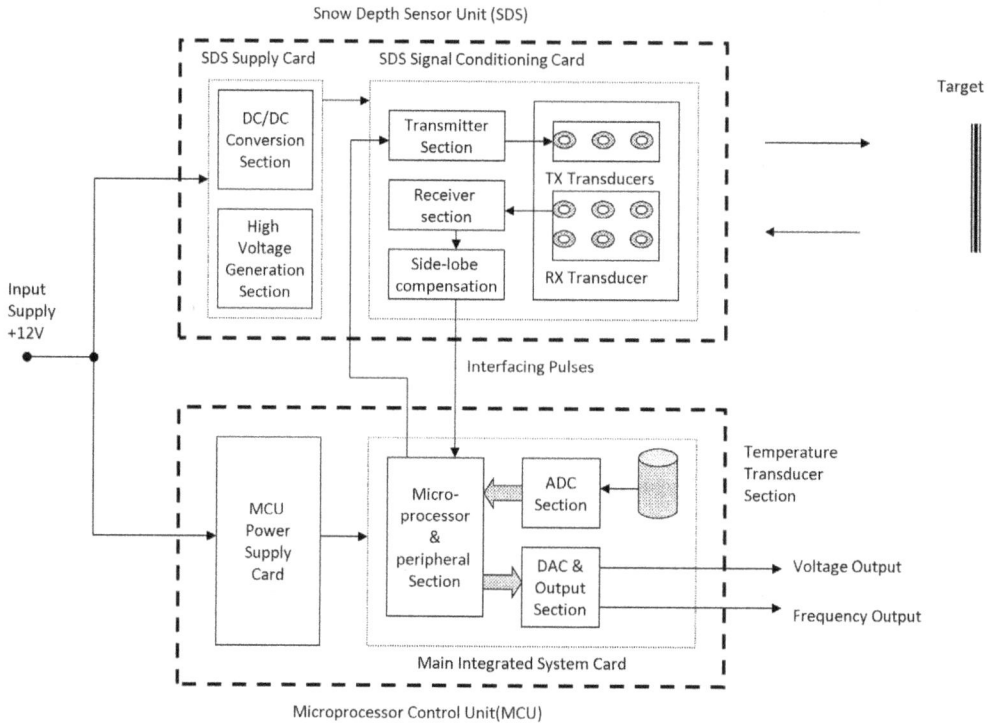

Fig [2]: *Detailed system block diagram of snow depth sensor*

4. DESIGN APPROACH

The design efforts have been made toward improving:
i)   extraction of 40 KHz signal from white noise
ii)   dynamic range
iii)   signal processing
iv)   wave-shaping and pulse-shaping
v)   preciseness and repeatability of distance measurements
vi)   the range of snow depth measurement
vii)   functionality in the low temperature range
viii)   universal standards of output formats

The design approach consists of modifying the directional response of the sensors, getting the effective transmission of ultrasonic energy and increasing the sensitivity of the receiver. This design approach is achieved by proper signal processing, noise removal and signal conditioning, amplification, and filtering backed by proper software-hardware co-design.

4.1   Design for Improved Directional Response

The first most thing is to choose proper ultrasonic generating transducers. For the present application, piezoelectric transducers operating at 40 KHz have been chosen. The higher frequency causes penetration in the porous snow surface and causes a lot of heating and absorption in the surface [7]. The range of measurement is very poor at higher frequencies. Frequency lower than 40 KHz is quite suitable as far as directional pattern and other losses are concerned, but then the accuracy of the measurement suffers. Further, the availability of lower frequency moisture proof completely sealed piezoelectric crystals working in −50 to +50°C temperature range is quite difficult. The experiments have been done to select the proper sensors. Refer to figure 3 which compares the range, accuracy, and penetration of an ultrasonic beam from piezoelectric sensors at different available frequencies [8]. Penetration is maximum at 50 KHz, and accuracy is best with 25 KHz. Considering the range, accuracy, and penetration inside the snow. 40 KHz crystal gives the optimum performance, which has been selected.

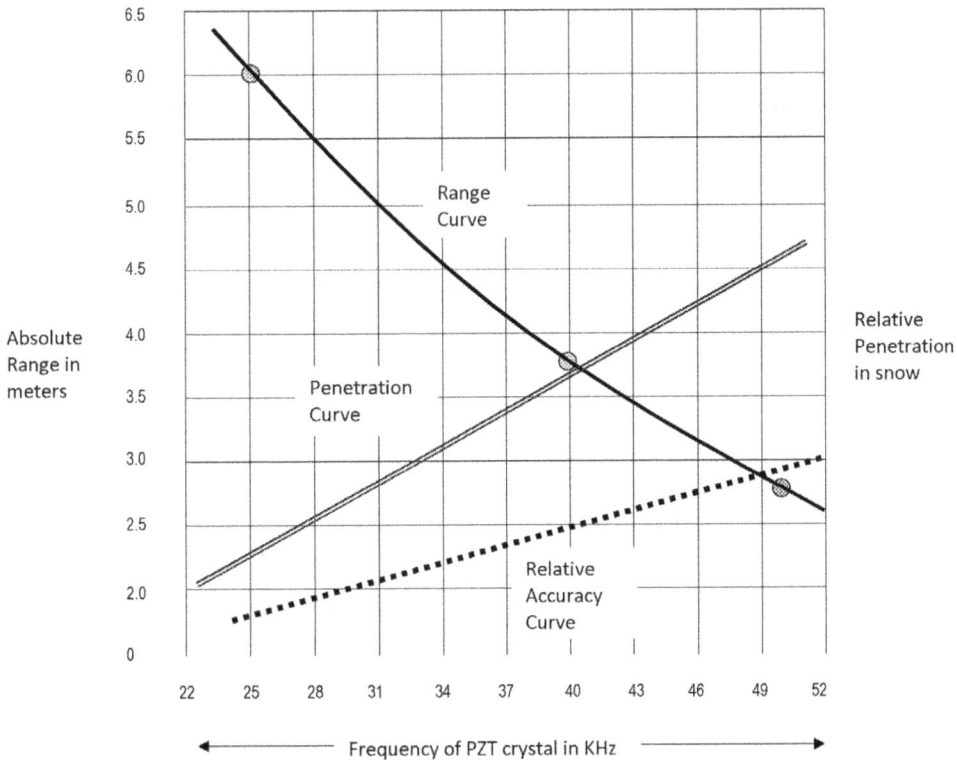

Fig [3]: *Graph of values of the range, penetration and relative accuracy of piezoelectric crystals as a function of frequency*

The Hall effect produces longitudinal mechanical waves along the axis of the crystals. The specifications of the Piezoelectric Crystal used are given in the table [1].

Table [1]: Specification of the PZT ultrasonic transducers

Parameter	Value
Model	MA40E6-7 or MA40E7S-1
Nominal frequency	40.0 kHz
Sound pressure level	> 108 dB at 40 KHz at 30 cm, 10 V_{rms}(Sine wave)
Sensitivity	> -82 dB at 40 KHz
Capacitance	2200 pF+ 20% at 1khz

Further, the mounting arrangement of these sensors is not conventional. Although the crystals used are reversible (can be used in transceiver mode), even then separate sensors for the transmitter and a receiver are used. Further, a one-to-one ratio of transmitting and receiving sensor has not been found suitable in the present application.

To deal with the non-smooth surface of snow and to counteract the scattering effects, the directional pattern of the transmitted ultrasonic beam has been modified using the ultrasonic transducer array arrangement. The beam pattern has been made relatively broader with large front zone area [9]. This design effort increases the possibility of getting even the scattered beam from the non-smooth surface of the snow. The configuration consists of separate 3x2 receiver array and 3x1 transmitter array isolated from each other mechanically and electrically. The configuration improves the directional response and beam width considerably, and optimum performance can be achieved by varying the geometrical parameters and the frequency of transmission. Refer to figure 4 which shows the directional response of the array transducers [10].

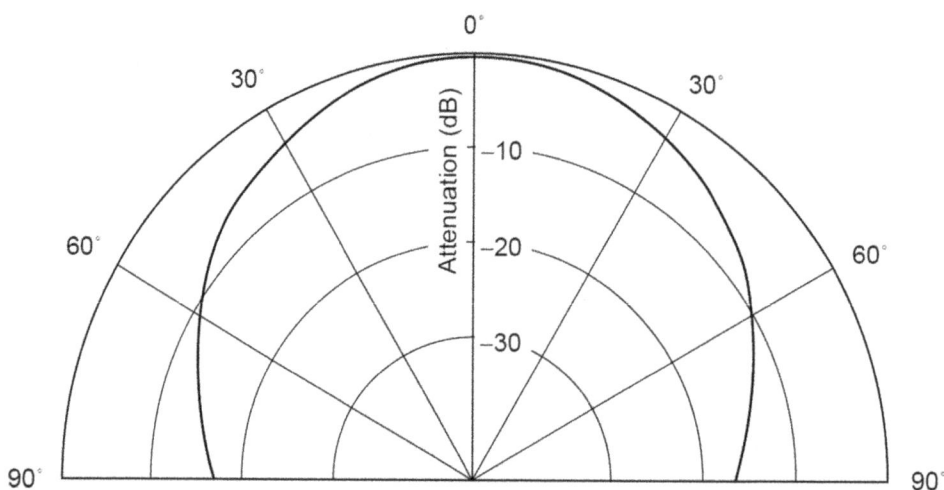

Fig [4]: *Directivity of MA40E7S-1 transmitting piezoelectric transducer in terms of sound*

pressure level measured in dB at 30cm with 40 KHz frequency. Typical directivity is 75°
symmetric in all directions. This results in a broader beam with a broad front zone area of the
beam (Source: Curtesy Murata Corp Japan) [19]

4.2   Design for Improved Transmission

The primary response of the snow depth sensor depends upon the amplitude of the transmitted signal. Greater the amplitude, better shall be the reflected echo. Normally piezoelectric ultrasonic transducers are operated at high voltages as high as 100 V. Experimentally it has been observed that an increase in transmitted amplitude has a very minimal effect on the response of the system. With an increase in 10% of transmitter amplitude voltage, only 1% improvement in the response is achieved. So, the transmitter voltage cannot be raised beyond a threshold, if raised at all, the improvement of response is minimal at the cost of heavy current consumption. 60 V has been chosen as optimal transmitter voltage after many experiments. The circuit diagram of the transmitter section is as shown in the figure 5.

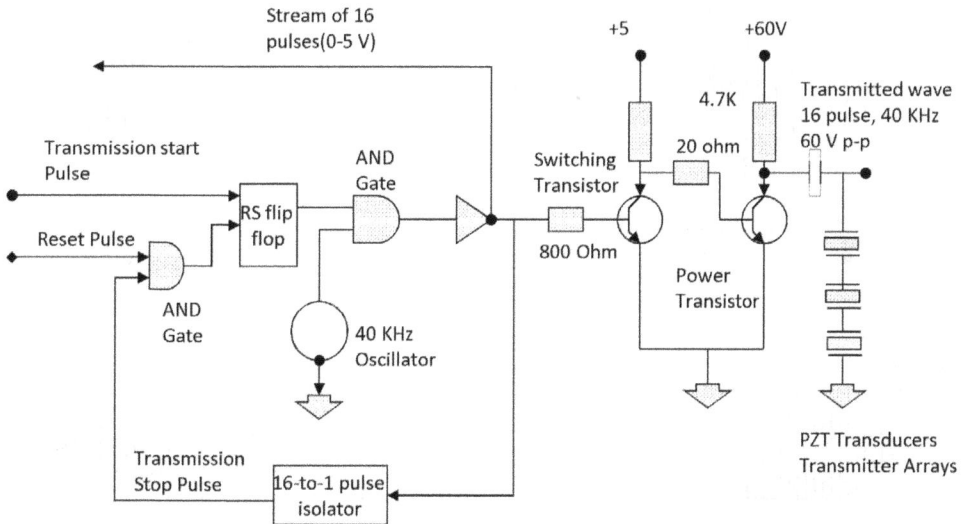

Fig [5]: *Transmitter section is a class-C amplifier operating at 60V, driven by a switching transistor. An oscillator and RS flip-flop combination logic ensure the passage of 16 pulses to the power amplifier and to a series array of transducers.*

A single pulse triggers the transmission section. Transmitter operates on 40 KHz frequency generated by 40 KHz oscillator. In the pulse-transit method, only a short burst of pulses is sent toward the target. The number of pulses chosen for transmission can be critical depending upon the application [11]. Since target here is quite an irregular surface with discontinuities, the number of pulses has to be more than the generally used. Experiments have been done to find out an optimal number of pulses for transmission. At the receiver side, the transmitted pulses make a modulated kind of envelope with a peak at the center of the envelope. The pulses at the extremes of this have very low amplitude. The 16-pulse pattern has been found to the most suitable pattern as a burst of an ultrasonic.

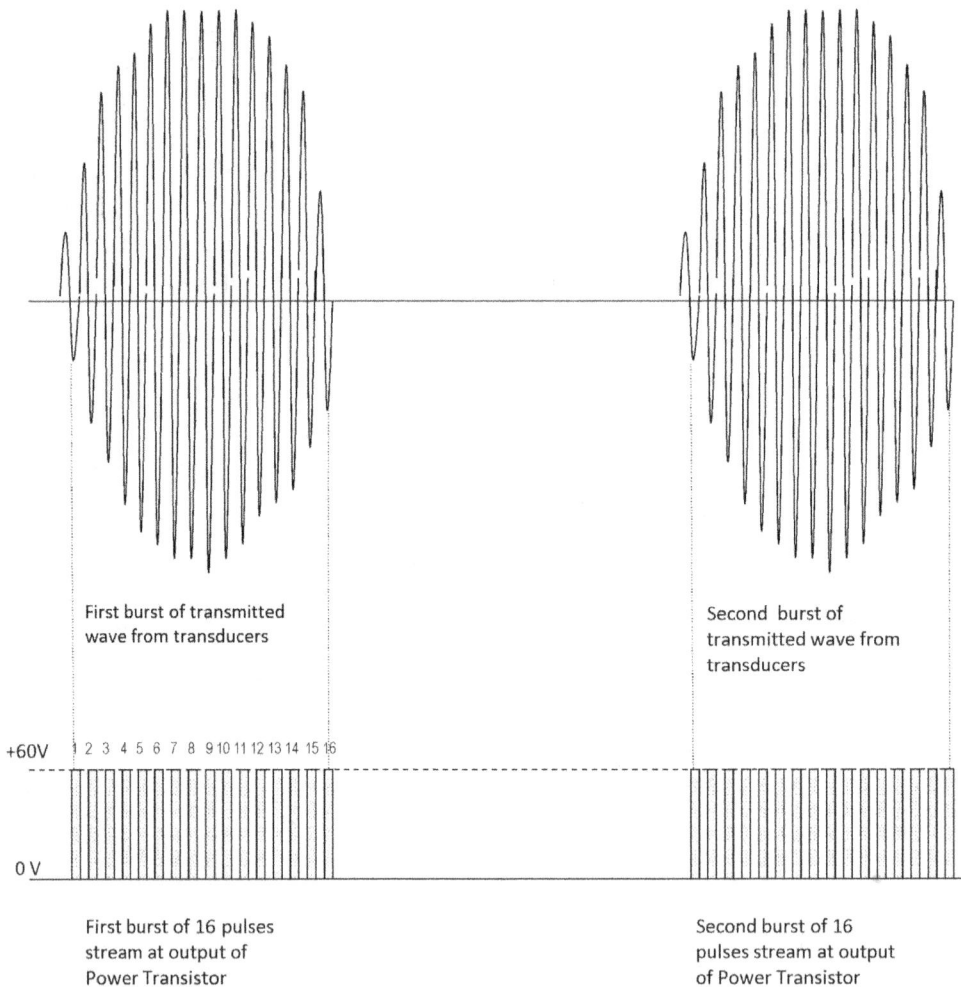

First burst of transmitted wave from transducers

Second burst of transmitted wave from transducers

+60V   1 2 3 4 5 6 7 8 9 10 11 12 13 14 15 16

0 V

First burst of 16 pulses stream at output of Power Transistor

Second burst of 16 pulses stream at output of Power Transistor

Fig [6a]: *A burst of 16 ideal square pulses of 60V triggered by output amplifier results in modulated mechanical waves at the transmitter transducer. The received signal also exhibits similar modulated characteristics.*

An R-S flip-flop in conjunction with the 16-pulse counter is used to transmit only 16 pulses to the base of the Transmitter Amplifier. The transistors are being operated in switching mode. The output stage is a class-C power amplifier operating at +60 V. The transmitting transducers are coupled to the output stage through a coupling capacitor. Precaution is taken not to apply direct DC voltage to the piezoelectric transducers for a long time. The duty cycle of 25 % is chosen to give a tolerance for the transition time of transistors and rise time of ultrasonic transducers.

Since the longitudinal mechanical waves generated from the piezoelectric sensor travels in the medium toward the target and received echo is reflected back and travels toward the sensor. Refer to figure 6a, which shows what is being transmitted. The transmission envelop is shown only for one PZT crystal while in actuality there are three envelops superimposing each other. This is clearer in figure 6b showing received echo waveform envelop.

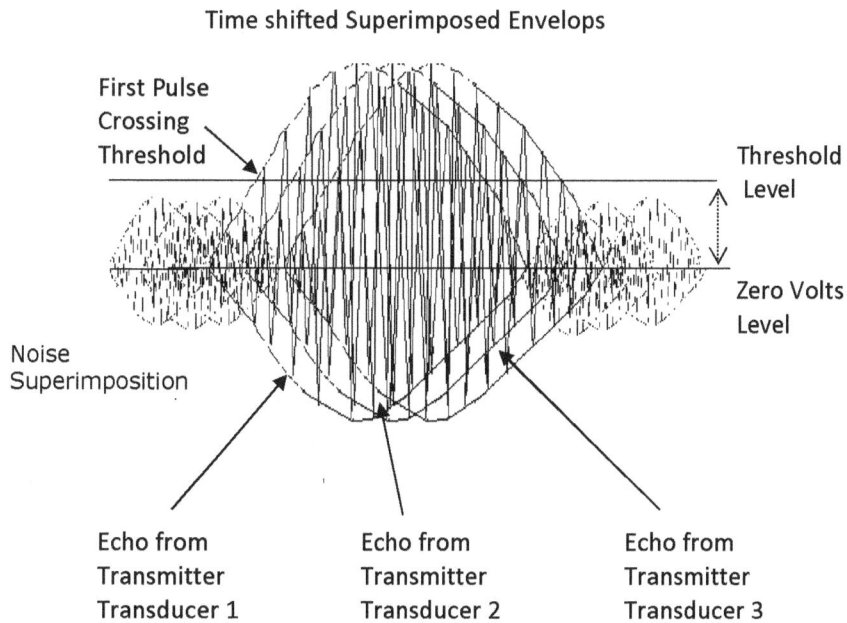

Time shifted Superimposed Envelops

First Pulse Crossing Threshold

Threshold Level

Zero Volts Level

Noise Superimposition

Echo from Transmitter Transducer 1

Echo from Transmitter Transducer 2

Echo from Transmitter Transducer 3

Fig [6b]: Superimposed time-shifted 3 transmitted wavefront envelops. Due to the capacitive effect, the three transmitters get triggered with a short delay in between. This results in a superimposed three envelop of pulses triggering in sequence

4.3   *Design for Improved Sensitivity at the Receiver Section*

The receiver section has been designed with extreme care because the whole performance of the system depends upon the fact that how accurately the reflected echo is detected by the system [12]. The reflected signal received at the output of the piezoelectric sensor is as minute as 1 mV in response to the 60V transmitted signal. This reflected echo is shadowed so heavily by the noise of higher amplitude that it is impossible to detect it without signal conditioning. The reflected echo is applied to the input amplifier stage which server the dual purpose of impedance matching and initial gain. Refer to figure 7 showing the receiver circuit diagram. Transistor self-bias with quotient point in the middle of the load line is used as input impedance matching stage. The signal is coupled to this stage through a coupling capacitor as the signal of interest is alternating at 40 KHz. Q-point selection is extremely important here, as no part of the signal should get trimmed, and the faithful amplified signal should be provided by this stage.

The gain provided by the input stage is 200. The amplified signal is still highly noise-ridden, noise spanning over the entire spectrum [13]. The output signal as shown in figure 8a is still in millivolts range. A high gain amplifier is used for enough amplification of the signal. An optimum gain of approximately 50 has been given at this stage. It is usually impossible to give very high gain at this stage because of Gain-Bandwidth product constraints. The useful signal bandwidth is approximately 50 kHz, which implies unity gain-bandwidth of approximately 2500 kHz. A video amplifier with UGB of 120 MHz has been used to amplify all the frequencies uniformly. The output signal now is of the order of 1 V, but still shadowed by the noise of higher amplitude. Experiments show that the high amplitude noise is well above or well below the center frequency of 40 KHz. As shown in figure 8 the reflected echo is still not reproduced faithfully.

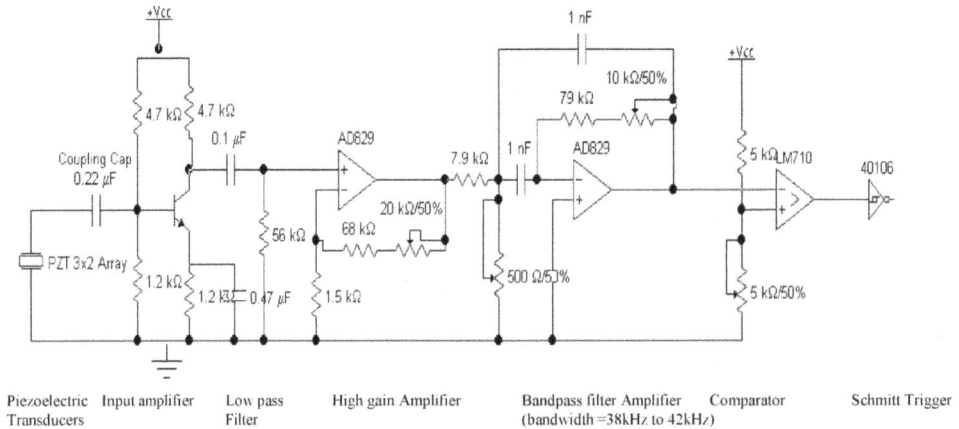

Fig [7]: *Block diagram of the receiver wave-shaping circuit. The circuit amplifies the weak received signal, compare it with a nominal threshold to eliminate the white noise and shape the resultant signal into square wave pulses*

A precision filter amplifier has been designed as a narrow band-pass filter with the sharp cut-offs at 38 kHz and 42 kHz with the center frequency of 40 KHz. The detection of the reflected echo and hence the sensitivity of the measurement depends very much upon the accuracy of the center frequency of this filter. This is a kind of notch band-pass filter, which should reproduce only 40 KHz frequency faithfully [14]. All other frequencies are filtered out. The response of this filter stage is shown in the figure 8c. The resultant echo is of very clear shape and amplitude. The number of pulses received in this echo is three times the number of transmitted pulses (because of three transmitters). Actually, the received echo consists of three envelopes shifted slightly in phase and each envelope having 16 pulses. The increased number of pulses gives better chances of triggering of the next circuit and hence the measurement and hence better sensitivity. The pulses received are conditioned and wave-shaped to drive the digital circuit, refer to figure 8e. The first pulse out of an incoming pulse stream pattern (figure 8f) is extracted as only this pulse corresponds to the arrival of the echo itself. The first pulse extracted from the transmitted stream has started a timing counter, and this first pulse of an incoming stream stops this counter.

The counts in the counter are a direct indication of flight time (time taken by the echo to travel to the target and return back to the sensor). This time is used in the computation of the distance of the target surface from the sensor as described at the beginning of this paper.

One important point observed is that despite the accurate filtering of noise, the residual noise is always present at the input of the sensors even when they are not receiving. This residual noise gets amplified and appears at the output of the final filter amplifier and is present all the time. The reflected echo waveform gets added to it. Generally, cross convolution techniques are employed to remove such residual noise. In the present case, an alternative method of comparison with some fixed threshold has been used. The average value of residual noise is checked out. A comparator circuit as shown in figure 7 is used to compare the filtered output to this threshold voltage [15]. If the reflected echo above this threshold, comparator gives the output in the form of pulses crossing the threshold mark. The comparator also wave-shape the bipolar sine signal into unipolar square pulses.

One important precaution in this design is setting gain and the threshold value of the circuit. A tradeoff point has been observed beyond which accuracy may get impaired if we further lower the threshold and increase the amplification to increase the system sensitivity.

Once we have obtained the pulse corresponding to the start of transmission and that corresponding to reception, we can calculate the flight time either using pure hardware or using the software. Both the methods have been implemented. Software in assembly language is a better choice as it gives the added advantage of calculation and system control together. The half the flight time, when multiplied to the velocity of sound at that temperature, gives the distance of the reflecting surface from the sensor. By subtracting this distance from the mounting height, we get the snow depth at that time.

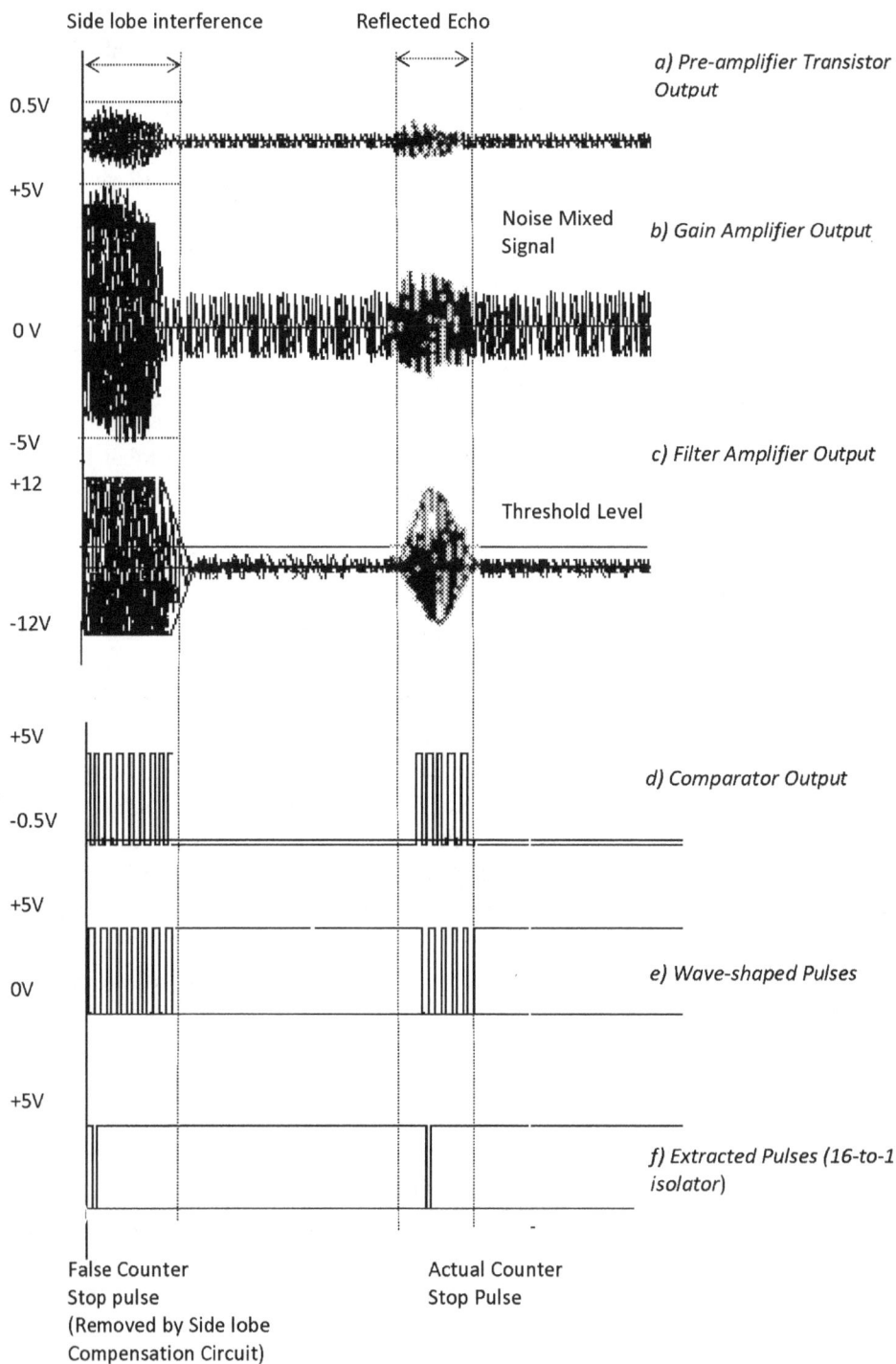

Fig [8]: a) Input signal at pre-amplifier transistors with low amplitude ridden in the mechanical noise b) Output signal from gain amplifier which amplifies the weak signal and amplifies the *background noise. The prominent signal level corresponds to the reflected echo. First saturated signal corresponds to either side-lobes or false trigger due to broader beam c) Output signal at Filter Amplifier output which cleans the signal and removes the noise. The reflected echo signal is enhanced for digitization. d) Pulse stream at comparator which removes the background residual noise and digitizes the inputs signal peaks above the threshold voltage e) Pulse stream at wave shaper output removes the negative side of the pulses and produces clean rectangular pulses from 0-5V f) Output of 16-to-1 pulse isolator circuit which produces one pulse against a stream of 16 pulses. This single pulse is used to stop the timing counter for measuring the transit time*

5. SYSTEM PERFORMANCE EVALUATION

The system has been designed to work not only for sampling-based measurements but also for the continuous round the clock functioning. The system has been implemented in a single mechanical unit for easy handling and installation. The mechanical chassis and design are an absolutely very important aspect of the overall engineering of the system. The system is expected to work round the clock in the heavily snowbound areas and continuously invaded by the harshest possible environment. The system still has to give the accurate reading of the snow depth in this type of hostile environment as well. The component failure and system resistance to moisture may hurt the performance of the system. As the system are to be installed at those places where frequent man visits and repairs etc. are impossible. So, the system has to undergo many tough tests. The three performance tests are used over this system:
i)   long-term stability testing
ii)   environmental testing
iii)   integrity testing

The system is sealed into a waterproof chassis and put into operation in a cold chamber at −40°C for 24 hours. These tests conform to the J55555 military specifications.

The long-term stability of the readings is checked by recording the distance shown by this system continuously for 24 hours. The fix target distance shown by this sensor is checked to be the same during these 24 hours.

The total system when interfaced with the automatic weather station or data collection platform is tested for accurate distance at all level of the moving target [16]. The cable-drop through which it is interfaced to the data-logger is chosen such a way that it gives minimum voltage drop over the length of the cable.

The overall performance and accuracy of the system also depend upon the components and the PCBs. Military specification JM38510/ JM883-B components have been used which can operate in the temperature range of − 60°C to +125°C. Gold plated PCB has been designed. The military-standard weatherproof connectors and cables have been used.

The software in conjunction with the hardware is used to improve the system performance. The accuracy of $\pm$ 1 cm with the resolution of 1 cm is achieved for the range of the 5 meters. Snow depth up to 5 m above the ground can be measured accurately with this system. System performance over all kind of porous and irregular surfaces has been checked out and has been found very reliable.

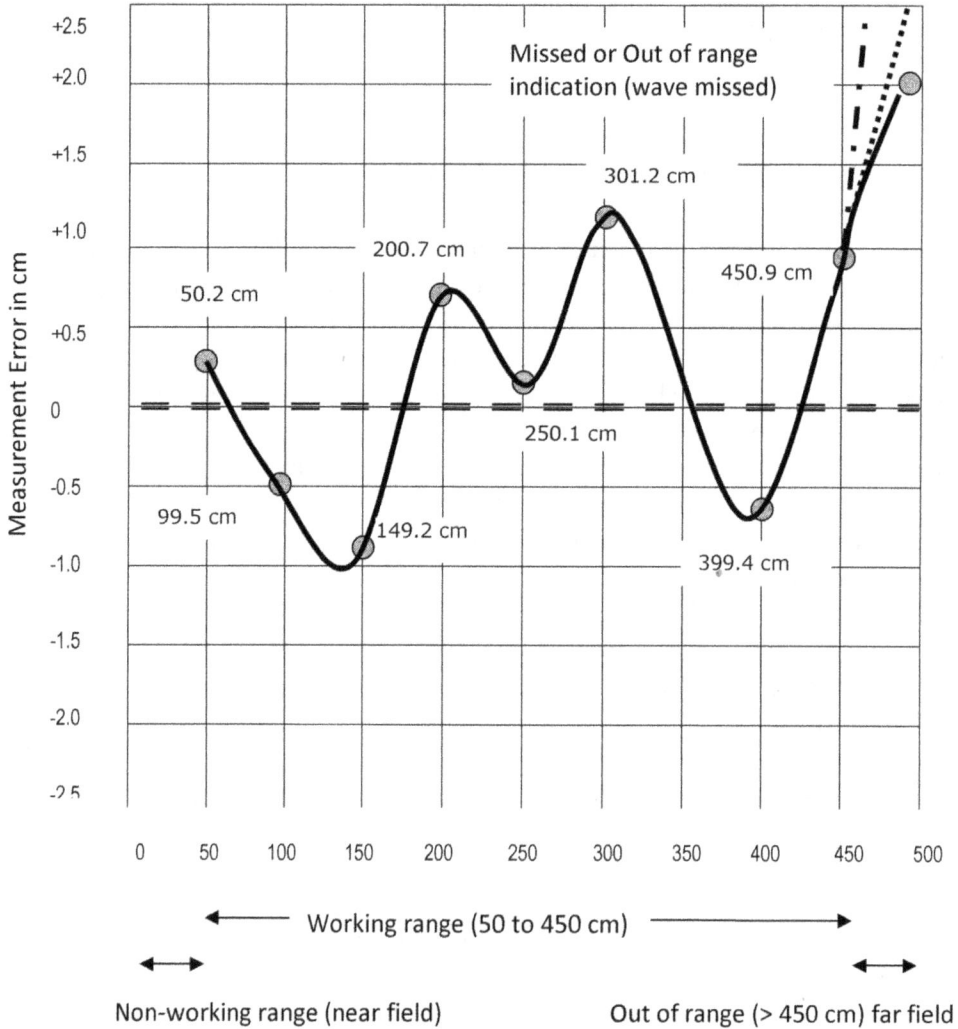

Fig [9]: System performance diagram showing error in cm against the zero-error axis at different points within the range. The system records distances between 50cm and 450 cm accurately. As it moves away from 450cm, the scattering effects come into the picture which may make the reflected beam to miss the receiver altogether. The accuracy is impaired in the far-field. Detection of the surface is inhibited up to 50cm in the near-field due to side-lobe compensation (the captions on the points indicate the measured absolute distances in cm)

System performance diagram in figure 9 shows the error in cm at different points in the range of the sensor. The wave is sometimes missed and not received back as an echo if the sensor is away from the target surface over 450 cm (far-field). It does indicate of out of range target surface through software programming. On the other hand, a near-field error comes in the picture due to side-lobe compensation and system do not detect any surface when it is near 50 cm to the sensor face. The side-lobe compensation circuit sacrifice near-field 50 cm measurement range, which practically is never used. The snow surface hardly reaches within 50 cm proximity of the sensor. The system accuracy is within ± 1 cm range.

6. DESIGN USED FOR COMPENSATION OF OBSERVED ERRORS AND IMPROVEMENT OF SYSTEM RELIABILITY

The major factor, which affects the system performance, has been the variations of the sound velocity with temperature. This compensation has to be provided. An integrated solid-state temperature sensor AD590 has been used along with its linear circuit, which reads the temperature with the resolution of 0.01°C. 12-bit ADC is used for digital conversion of temperature. Extensive Software computations have been carried out to compensate the effects of the temperature over the velocity of the sound.

While checking the system performance, it has been observed that the two streams of the received pulses are obtained, first one with a large amplitude. The first stream comes immediately after the transmission and another stream arrives late. The first stream of pulses is corresponding to the side-lobes of the transmitted beams, produced because of the near-field interference [17]. This first pulse will produce a counter stop pulse, which shall stop the timing counter prematurely. This effect is called side-looking, and system presumes as if an object is lying near the sensor. The first stream of pulses is blocked by hardware and software implementation. The figure 10 shows the circuit for blocking the first stream of pulses till the total width of the first stream so the sensor responds to actual echo only.

The voltage level at the input of OR gate is made high till the expected time of first steam. The expected time has been around 2300 microseconds. A safe margin is given and pulses till the time 2560 microsecond are blocked. After that, any stop pulse arriving shall be passed through and the stop the counter. In software, any reading below counts corresponding to this time are rejected, and measurement is termed false, and the cycle is again repeated. This is near-field compensation.

At the beginning of the discussion, it has been highlighted that if the snow surface is much beyond the range of the sensor, then the reflected signal echo will be much below a threshold level and it will not be detected. So, system timing count shall keep on increasing with no intelligent sensing of missed echo. The situation shall be same because of the severe scattering echo beam arrive the sensor at such an angle that it completely misses the sensor. Here software checks for the maximum expected counts. If current counts cross the maximum permissible counts, the system takes the situation as the missed beam and repeat the measurement cycle until it gets the counts in the permissible accurate range. Even when the surface is beyond its range, during some measurement cycle, it has been found that incident beam strikes the surface at such an angle that the scattered beams make high amplitude interference pattern at the receiver sensor. This caused detectable echo amplitude even when the surface was beyond the range of the system. This far-field compensation enhances the system performance and makes it intelligent enough to detect the possibility of the surface in the path.

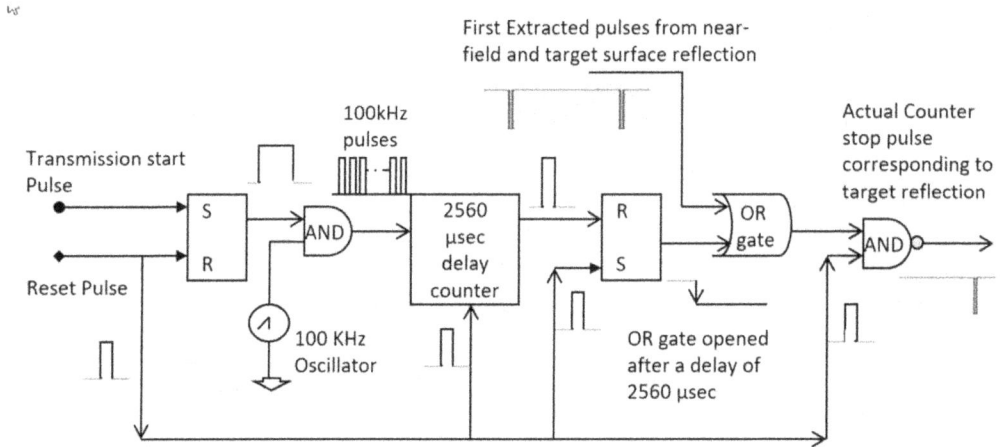

Fig [10]: *Near-field compensation circuit to stop the system counter to avoid false triggering of receiver circuit due to broad beam pattern from the transmitter and/or side grating lobes.*

7. CONCLUSION

The experimental results have been very favorable. A considerable reflected signal is received through arrayed receivers, which is detected with the help of the electronics design. The sensitivity, accuracy, long-term stability and range of the system has been enhanced using these design techniques. The snow surface distance over 5 meters can be fairly detected despite the nuneven, rough, porous snow surface which normally does gives problems of scattering and absorption of energy into the snow. The directional patterns have been made broader to offset the effects of non-smoothness of the surface. The reliability of snow depth parameter, which is measured by this snow depth sensor, is extremely important for the modeling of snow avalanche forecast and other related studies [18, 19]. This snow depth data is recorded after suitable intervals by the data collection plate form and keep track of deposited snow and snow-melt.

ACKNOWLEDGMENTS

Dr. B.K. Sharma, Head of Dept., Geo-Scientific Instruments Division, CSIO, Chandigarh

Swaranjit Singh, Senior Technical Officer, Geo-Scientific Instruments Division, CSIO, Chandigarh

REFERENCES

[1] Shamshi, M.A., Attri, R.K., Sharma,V.P., Snow Pack Temperature sensor, Proceedings of National Conference on Sensors and Transducers, pg. 180-189, 1996.

[2] Mellor, M., Engineering properties of snow, Journal of Glaciology, volume 19, pg 15-66,1977

[3] Satish Kumar et al, Snow Depth Senor, Proceeding of national symposium on Sensors and Transducers, 1996.

[4] Gooberman G.L., Pulse Techniques, Ultrasonic Techniques in Biology and medicine, Illiffe books Ltd, London, 1967 pp 87-128.

[5] Balantine D.S et al, Acoustic Wave Sensors-theory, design and physico-chemical applications, Academic Press, 1997

[6] Busch I, Huxson E, Encyclopaedia of applied Physics, VCH Publishers, vol 1, pp 63-88, 1991

[7] Ensminger D, Ultrasonic – the low and high intensity applications, Marcel Decker Inc, New York, 1973

[8] Krautkramer L and Krautkramer H., Ultrasonic Testing of materials, Berline: Springer, 1969.

[9] Encyclopedia of Science and Technology, Vol 12, pp 662-664, Mcgraw Hill, 1982.

[10] Hudson J.E, Adaptive Array Principles, Peter Peregrinus , London 1981

[11] Silk M.G., Ultrasonic transducers for non-destructive testing, Adam hilger Ltd, Bristol, 1984

[12] Zhang D, Crean G.M., Non-linear acoustic properties of high purity quartz as function of temperature, Proceedings of Develop,ment in Acoustic and Ultrasonic, IOP Physical Acoustic Group , Leeds, 24-25 Sept 1991, pp 219-224

[13] Landee, R.W. et al, Electronics Designer's handbook, Mcgraw Hill , 1957

[14] Raman S et al, Processing of Ultrasonic signal for transducer characterisation and for improving signal to noise ratio, Indian Journal of Technology, Vol 31, Nov 1993, pp 774-776

[15] Lawson, J.L. and Uhlenbeck, G.E., Threshold Signals, Mcgraw Hill, 1948 pp 211

[16] Ganju A, Snow cover model, Proceedings of SNOWSYPM-94, pg. 221-226, sept 1994

[17] Szilard J, Ultrasonic testing – non-conventional testing techniques, Willey, NY, 1982

[18] Jordan, R.. A one dimensional model for snow cover, CRREL, Special report 1991

[19] Yamazaki, Kondo, T. J., Sakuraoka,T. and Nakamura, T., A one dimensional model of evolution of snow cover characteristics, Annual of Glaciology, 18, pg 22-26,1993,

[20] Attri Raman K., Sharma B.K., Shamshi M.A., Practical Design Considerations for Signal Conditioning Unit Interfaced with multi-point Snow Temperature Recording System. IETE Technical Review, Vol 17, No 9, pp 351-61 Nov-Dec 2000

[21] Papadakis E.P., Physical acoustic principles and methods, (W.P. Mason , ed.), vol 4 B, Academic, NY, 1968

[22] Hall, Derothy K., Remote sensing of Ice and Snow, London, Chapman and Hall Publications, 1985,

[23] Attri, RK 2018/2005, 'Design of a Reliable Remote Surface Detector Based on Ultrasonic Pulse- Transit Technique to Detect Uneven & Non-Smooth Porous Snow Surfaces,' R.Attri Instrumentation Design Series (Snow Hydrology), Paper No. 1, *Research and Design of Snow Hydrology Sensors and Instrumentation*, 2nd edn., pp. 1-23, Speed To Proficiency Research: S2Pro©, Singapore.

[24] Attri, RK 2018/1999, 'Implementation of Linear Array of Ultrasonic Transmitter-Receiver Transducers for Detection of Non-Smooth Porous Surface,' R.Attri Instrumentation Design Series (Snow Hydrology), Paper No. 3, *Research and Design of Snow Hydrology Sensors and Instrumentation*, 2nd edn., pp. 43-57, Speed To Proficiency Research: S2Pro©, Singapore.

IMPLEMENTATION OF LINEAR ARRAY OF ULTRASONIC TRANSMITTER-RECEIVER TRANSDUCERS FOR DETECTION OF NON-SMOOTH POROUS SURFACE

RAMAN K. ATTRI

EX-SCIENTIST,
CENTRAL SCIENTIFIC INSTRUMENTS ORGANIZATION INDIA

Manuscript originally written July 1999

Abstract - Level measurements, thickness measurement or remote surface detection using Ultrasonic Pulse-transit Method require that the target surface be at 90O to the incident beam, so that reflected beam comes back at 180O angel to effectively use this method. This is perfectly true in case of flat, solid surface at a right angle to the incident beam. However, surface irregularities of a porous, non-smooth, uneven material such as snow cause penetration of wave into the surface, absorption of the incident energy, the scatter of energy in many directions and further of reflected making it difficult to detect the reflected echo. Such a received reflected echo is very low in amplitude and is heavily noise-ridden. The successful surface detection with excellent repeatability and accuracy with no false measurement requires a combination of physical acoustic, mechanical, hardware and software techniques together. In this paper, we discuss the suitable physical, mechanical, electronics design to physically implement the theory of Arrayed Ultrasonic transducers to shape up the directional response, beam width and avoid interference to improve the chances of proper and sufficient reflection from the non-smooth, highly porous, uneven, non-planar irregular surface.

1. *INTRODUCTION*

Remote Surface Detection using Ultrasonic Pulse-Transit (Pulse-echo) method is one of the popular methods for detecting the target surface, estimating its distance accurately and its 3-D imaging [1]. Such kind of remote surface detector (RSDs) have numerous applications in industrial, military and physical instrumentation environments. Most such methods are used for solid, smooth and possibly a flat surface or a target having a smooth surface. In such cases the chances of reflected energy back to the source are high, and the target is detected accurately.

However, the issues do not remain simple when the same method of Ultrasonic Pulse-transit is used for detection of targets with non-smooth surfaces, level detection where the surface is uneven and for detection on non-sold porous surfaces. Such detection of non-solid, porous, uneven, non-smooth, irregular surface requires a lot many considerations in the design of such RSDs. There are numerous applications, where the non-smooth, uneven, porous, irregular surface detection or target detection is an important application. For example, the snow surface detection is a very important application of such RSDs to determine the thickness of the snow layers by finding how much the current surface level is above the ground level. This requires detection of snow surface detection distance from the sensor [2]. This information is used for hydrological studies such as the forecasting model for snow avalanche release, river run-off water, glacier sliding and related phenomenon in the mountain areas and planes nearby [3, 4]. The fresh snow is extremely porous non-smooth, and the irregular surface where using remote sensing of snow using ultrasonic beams has its own problems [3, 5].

Another example is the automated chemical plant where either liquid or some powder compounds are being filled in big tanks, and upper surface level of the compound w.r.t to the base of the tank is detected by an ultrasonic sensor so as to stop the filling jet at the right time to avoid the spilling. Liquid or oil level in the tank is one of such application employing ultrasonic techniques to surface level detection [6]. This application is a typical case of non-smooth porous and continuously growing surface. Another application may be the detection of a porous surface of the sand. In such applications, the Ultrasonic Pulse-transit Method is generally employed, but the accuracy requirements make it a critical application [1, 7].

In this paper, we will discuss the theory behind the arrayed transducers in series and its physical implementation to produce improvement in detection of such non-smooth irregular surface of a porous material (typically snow). This method of minimizing the effects of losses in the material because of its irregularities and increasing the reliability of the signal received will be discussed keeping the electronics design and software algorithm out of the scope of this paper. The scope has been limited to issues and their solutions through physical, mechanical or electronics design only.

1.1   *Pulse-transit (Pulse-echo) Method of Level Detection*

The level detector is one special case of the remote surface detector which inherently detects the surface of the material or liquid whose level with respect to the base is to be found out. The ultrasonic based level detectors are quite popular in oil tank level determination applications [6]. The concept used in level detectors or surface detector or depth sensors is same. From Electronics viewpoint, this depth sensor system is a standalone system consisting of ultrasonic transmitting transducers, ultrasonic receiver transducers with a signal conditioner circuit and software-controlled hardware designed around a microprocessor. The level or the surface is detected using pulse-transit method [1]. A short burst of ultrasonic pulses is transmitted by the piezoelectric transducer transmitter, which is mounted on a pole with the sensors facing vertically downward toward the surface of interest. Refer to figure 1. The system is usually having separate or dual-purpose ultrasonic transducers fitted in its hood, facing the surface. The transmitter section works under the control of the microprocessor and is driven by high voltage power transistor giving out pulses in the ultrasonic frequency range of 40 KHz at around 60V to 100V amplitude. The frequency and voltage levels are application dependent and depend upon the nature of the surface being detected. Mechanical waves generated by transmitting ultrasonic transducers travels toward the target surface. The transmission is done as a short burst of ultrasonic pulses.

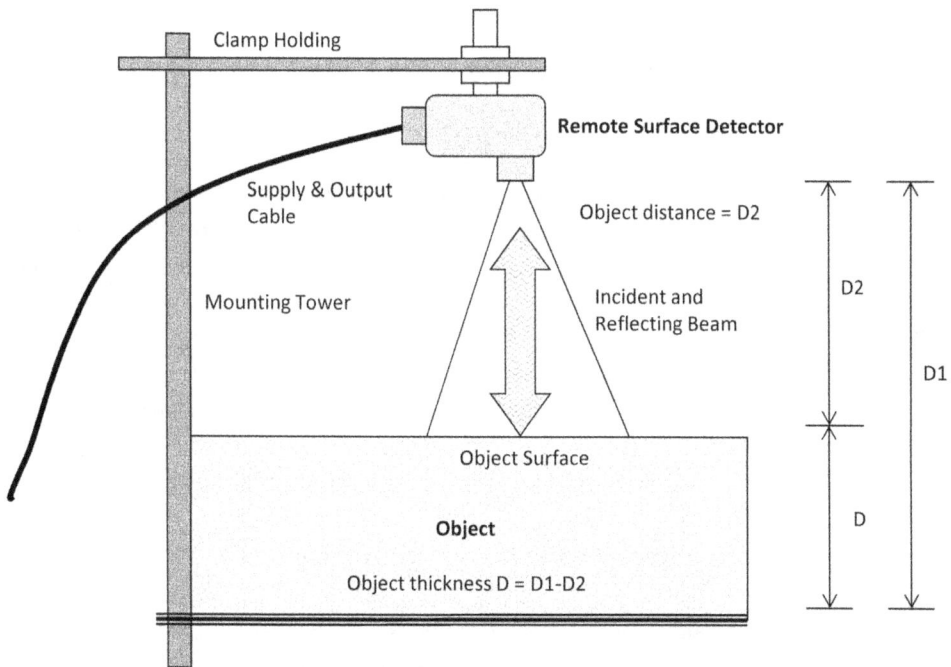

Fig [1]: Remote surface detector mounting arrangement

The transmitted beam strikes the surface of the target. Some of the energy gets reflected back and is received by the receiver. The receiver is a set of ultrasonic transducers mounted adjacent to the transmitter transducers. Sometimes, the same transducers act as a transmitter and a receiver. The receiver converts the received mechanical energy in an electrical signal forming a voltage echo signal in the signal conditioner [1].

The microprocessor reads the time lag between transmitted pulse stream and received pulse stream. The time of travel between the transmission and the reception of the reflected pulses (more precisely echo) is computed which gives the distance of the upper surface from the sensor as per distance-velocity equation. This distance is employed for determining the distance of the object remotely.

The other use of this distance is finding the net thickness of the material if the base distance (a distance of the sensor from the ground, i.e. sensor installation height) is already known. By subtracting the measured surface distance from the installation height, the surface level or the distance of the surface from the ground or base is obtained. This method is called the pulse-transit method since a short burst of pulses of ultrasonic energy is transmitted toward the surface and time of transit between transmission and reception is found [1]. This method is also called Pulse-echo method which means listening to the reflected echo of the transmitted beam [7].

The pulse-transit method can be used with sound waves, light waves or radio waves. However ultrasonic is most popularly used due to the nominal accuracy required in these applications.

2. PROBLEMS DUE TO ENERGY LOSSES IN NON-SMOOTH POROUS MATERIAL

The overall performance and reliability of the depth sensor based on this method depend upon the ability of the system to detect the reflected echo of the transmitted signal [1]. Further, the system performance can be increased only if one can enhance the reflected echo to a sufficiently detectable level.

This simple method of finding the thickness, depth or level is not as simple when applied to the snow or any other similar porous material surface detection. There are many factors like losses in the material and irregularities of the surface which govern the range and reliability of the depth sensor [8, 9]. Some of the critical problems associated with pulse-transit method when used for the non-smooth porous surface are as follows:

2.1 Attenuation of Received Echo Due to Divergence

One very first phenomenon that happens with the ultrasonic beam in all the applications requiring the long range is the divergence of the beam. Additional allowance has to be made for reduction of amplitude due to the divergence of beam [10]. The divergence of the beam determines the range in case of materials, which can readily be penetrated. As the distance increases the beam becomes less strong, and at a very long distance, the beam eventually dies out. The coefficient of attenuation is directly proportional to frequency. Attempts need to be made to avoid attenuation due to the distance by using lower frequencies. However, using lower frequency has another disadvantage of reduced accuracy in the range determination of the surface. A suitable tradeoff is required.

2.2   Scattering of Energy at Rough Surface Grain Boundaries

In the specific application of depth measuring, the shape and the roughness of the surface is of decisive importance. These factors often limit the sensitivity of the depth sensor. If we take the example of snow surface, a roughness of over 1/10 of wavelength impairs the coupling markedly [4]. The rough surfaces make the ultrasonic beam to become diffused and scatter in all directions, as shown in figure 2.

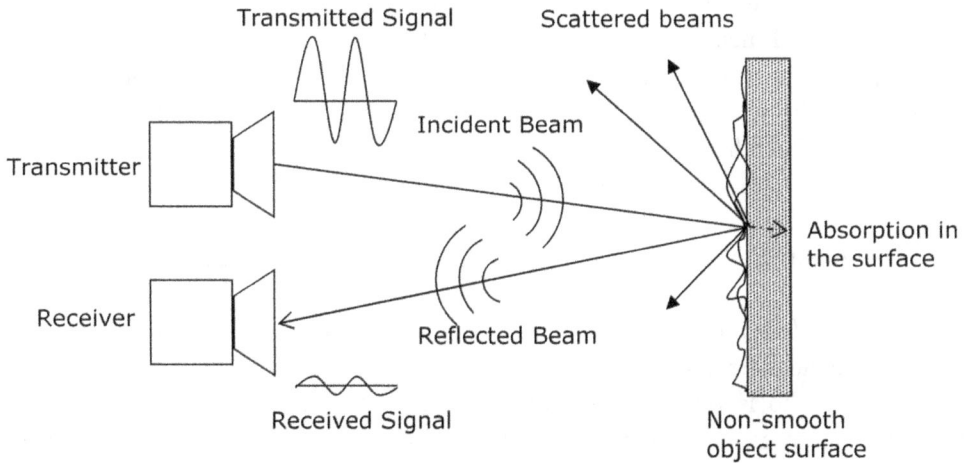

Fig [2]: Scatter and absorption of ultrasonic wave in a non-smooth porous surface

The major loss in reflected energy is the scattering of the energy at the grain boundaries. For non-smooth surface like snow, the grains are not strictly homogeneous and contain boundaries on which acoustic impedance changes abruptly because of change in density [4]. The average grain size of snow is 2 mm to 5 mm which is coarse as compared to the wavelength being used and as a result, scatter will take place as the splitting of the incident wave into various reflected and transmitted waves at the oblique boundary. This process keeps repeating at every grain boundary. So, scattering is a severe problem in case of snow, which reduces the effective energy getting reflected toward the sensor, and hence reducing the amplitude of reflected echo. The porosity of snow, where crystallites of different structure and composition are present, is also one of the similar flaws. The worst of all is that sometimes the reflected echo is not received at all. The reflected beam reflects at such an angle that it completely misses the receiver sensors and system remains in wait mode.

2.3   *Absorption of Transmitted Energy by the Porous Surface*

The porosity of such surfaces such as snow causes a large amount of ultrasonic energy to get absorbed in the snow, so only a small part of it gets reflected back. The phenomenon is depicted in figure 2. This absorption is another reason for weak reflected signal [10]. It is the conversion of sound energy into heat due to the oscillation of particles. The absorption increases as the frequency of the incident wave. To counteract this effect transmitter voltage and amplification has to be increased. Stepping up transmitter voltage or the amplification cannot counteract the much more awkward disturbances caused by scattering, which not only reduces the height of echo from both the flaw and the back-wall but also produces numerous echoes with different transit times, in which true echoes may get lost. Absorption and scattering both can be reduced by lowering the frequency of the transmitted pulses, for which again there is a limit [10]. The received echo is very weak which need considerable amplification [12, 13]. The worst of all is that the ultrasonic beam is striking the non-smooth surface, which causes the scatter, and the missing of a reflected beam. The system has to be very sensitive to detect the weak reflected echo. The reflected echo is around 0.1 mV in response to the transmitted wave of 100 V.

2.4   *Inherent Errors Associated with an Ultrasonic Beam*

Along with design constraints, some errors are also encountered in the system. It pertains to the near-field interference and grating lobes caused by this interference [9, 14]. The reflected echo is received because of these grating lobes instead of the actual reflection, thereby causing the effect of side-looking, implying as if the target is lying in the immediate proximity of the transducers [9, 15]. In the directional pattern, the side-lobes are shown in the figure 3. These grating lobes induce a signal in the nearest receiver situated on the side of these lobes; hence it is taken as a reflected signal by the unit, thus giving wrong distance reading as if the target surface is very near to the sensor hood. These side-lobes also depend upon transducer diameters and wavelength ratio [14, 15]. One obvious solution could be to modify the dimensions of the transducer, which is impossible as these sensors come in pre-fabricated form. However, by changing the geometry of multiple transducers in an array could be one solution as suggested by array theory to reshape the directional response and also some implementation in hardware to provide such near-field compensation, so that side-lobe is not detected at all.

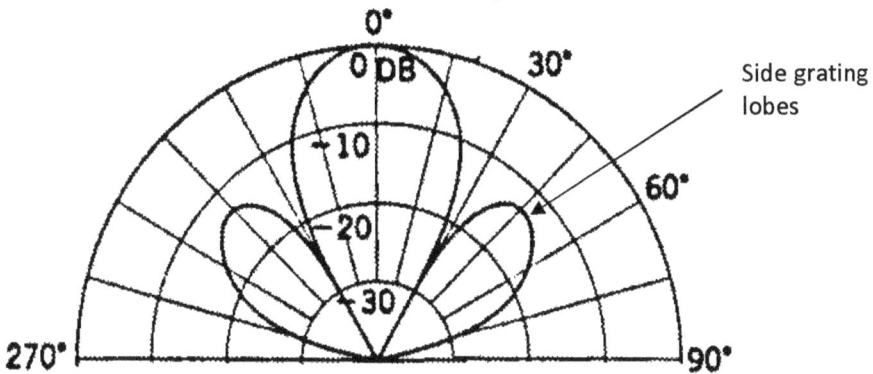

Fig [3]: Side-lobes in the directional pattern of ultrasonic transducers [14, 15]

To sum up, such critically irregular and non-smooth porous surface (e.g. Snow surface) causes penetration of a wave of the incident wave into the surface, absorption in the surface and scattering around which results in either missed reflected wave or a very low amplitude highly noise-ridden reflected signal [16]. These have to take into consideration while designing the system. It has been found that the echo reflected from non-smooth irregular snow surface is of such small amplitude which is difficult to detect. The received echo is very weak which needs considerable amplification. The worst of all is that the ultrasonic beam misses reaching the receiver and system keep waiting for the reflected beam.

3. DESIGN SOLUTIONS

The phenomenon of divergence, absorption, and scattering should be dealt with simultaneously through unified design techniques. Three obvious options have been:
   i)   Increase the transmitted voltage.
   ii)   Reduce the operating transmitted frequency.
   iii)   Increase the receiver amplification.

Experimentally it has been found that an increase in transmitted voltage amplitude does have some improvement of results, but this improvement gets saturated after a certain point, and no amount of increase in transmitted voltage make any effect. Using lower frequency does have some positive effects of increasing the range and reducing the losses due to scattering and absorption, but there is a limit on lower frequency ultrasonic transducers available. It further affects the range accuracy badly. Further, the increase in amplification of received echo will only increase the noise beyond the threshold level.

So, these obvious design approaches could not be used beyond a particular limit. However, we optimized the transmitter voltage by keeping the current consumption lower and the response at maximum. A relatively low frequency of 40 KHz was selected without sacrificing the accuracy much. A nominal gain and suitable threshold were provided at the receiver to limit the noise within the threshold limit so as not to give a false reading.

However, all these solutions could improve the reliability of the system, but scattering, absorption, and chances of altogether missed wave were still there. Besides normal electronics signal conditioning requirements, the optimal design had to fulfill these requirements:

i)   A broader surface area of the receiver
ii)   Broader beam width
iii)   Higher range (min 4 meters)
iv)   Accuracy of ± 1 cm in distance detection
v)   Lesser absorption in the surface
vi)   Increased amplitude of superimposed receiving pulses
vii)   Effective compensation of the scattering
viii) Effective cancellation of side grating lobes

We found that the theory of the planar array technique was quite suitable to solve most of our requirements effectively. In planar array technique, some transducer elements are connected in series to form an array and mounted in a plane [7]. The resultant directional pattern of the array is a mathematical function of the directional pattern of the individual transducer. The desired pattern, which increases the range and the sensitivity, is obtained by controlling the factors like several transducer elements, diameters of transducers, frequency, geometry, and inter-element spacing, etc. [15, 17]. By using a proper number of transducer elements, spacing and geometry, using this array of transducers, the directional pattern could be made a little broader to increase the chances of reception after selecting the diameter of the pre-fabricated transducer and selecting the optimal frequency.

4. PHYSICAL IMPLEMENTATION OF ULTRASONIC TRANSDUCER ARRAYS

Although the crystals used are reversible (can be used in transceiver mode), even then separate sensors for the transmitter and a receiver are used. The further one-to-one ratio of transmitting and receiving sensor has not been found suitable in the present application. A separate set of transducers have been used in the transmitter and receiver in the array mode. A different arrangement of arrays has been used for receiver and transmitter. The transmitted ultrasonic beam directional pattern has been modified using this ultrasonic transducer array arrangement to deal with the non-smooth surface of snow and to counteract the scattering effects. The shape of these individual elements, the spacing between the individual transmitting transducers in x-axis and y-axis, mounting geometry, i.e. whether in rectangle, square or triangle, etc. drastically shape up the directional pattern, beam width, and side-lobes. [14, 15, 16, 17].

To achieve the right kind of directional pattern we either need extensive mathematical computations to find the optimal geometry or to perform the extensive experiments to select the right mounting geometry. In our case, we performed the experiments keeping the theory of an array as the basis. The purpose in our case was to make the beam pattern relatively broader with large front zone area [17]. This design effort increases the possibility of getting even the scattered beam from the non-smooth surface of the snow.

Based on mathematically model of array theory and experimentation on it, for transmitting array we chose 3 x 1 line-geometry and for receiver array we chose 3 x 3 square geometry, mounted on the same plate in a planar fashion with same inter-element spacing between the elements for transmitter and receiver array. It is shown in figure 4.

The transmitter is a line geometry with transmitting array of 3 x 1 (i.e. 3 transducers mounted in a row) having a spacing between them equal to a little less than the half wavelength. Refer to figure 5a.

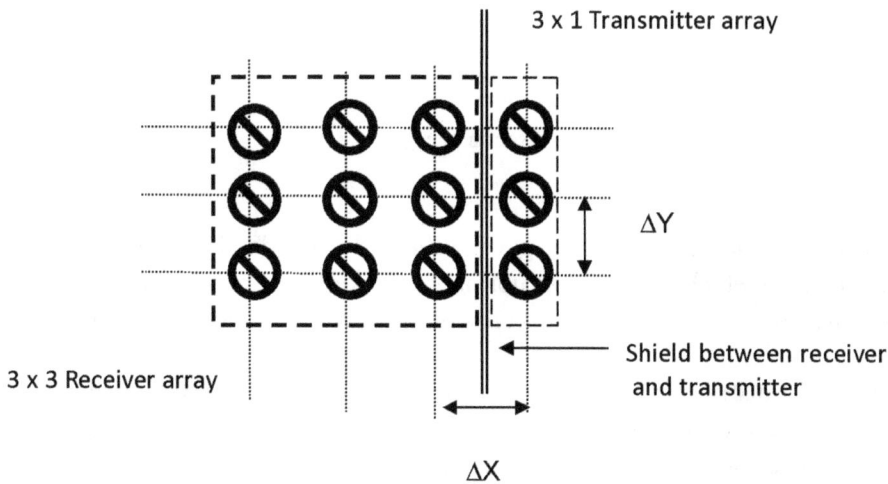

Fig [4]: Transmitter and receiver array arrangement

The transducers are connected in an array. There are two options of connecting the array: one is putting elements in parallel, and one is putting the array in the series [13, 17]. In our case, we have chosen the latter approach. It has given a big design advantage. Putting transmitter array in series, one shown in figure 5a ensured that each of the three transducers is activated with little delay in between, so the result is the transmission of three envelopes of bursts. This short delay makes the right superimposition at the rough surface and chances of reflection, from any of the envelops, is strengthened. The series connection at transmitter adds up the directional response of the three individual crystals, and resultant transmitted beam pattern is wide and directional enough to gives a beam focused on non-smooth snow surface covering a good amount of surface area so that at least form one point out of the covered area, there is a good possibility of reflected signal reaching toward the receiver array.

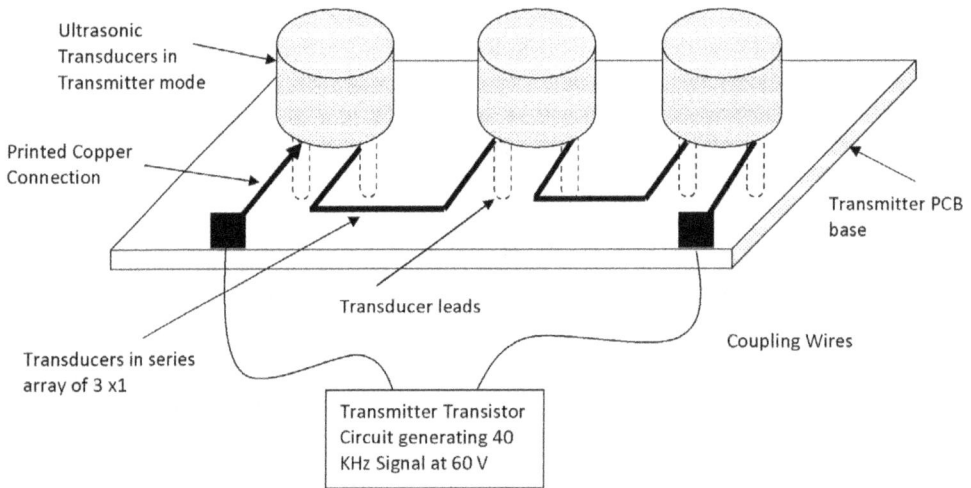

Fig [5a]: Transmitting transducer in a series array

The result was that the array gave sufficiently broader beam width with beam expansion of 60° which is double the normal beam width. Refer to figure 11 which shows the improved & wider directional response of the array transducers (explained in a later section). [15, 17, 18]

The power distribution of broader beam width was compensated by increasing the power supply to the transducers from 30V DC to 60V DC. With this configuration, the beam width becomes broader, so it travels as a broader beam rather than a straight-line thin beam toward the surface. Even if the surface is uneven, there are chances that out of a bunch of integrated beams some components will be reflected toward the receiver. Optimum performance can be achieved by varying the geometrical parameters further and the frequency of transmission, if we have an option.

The receiver array response was also optimized by using the rectangular geometry consisting of array configuration with total 9 transducers mounted in a square array of 3x3 receiver array isolated mechanically and electrically by a shield for proper decoupling of interference from transmitting array, as shown in figure 4 [17]. Wider receiver surface area gives the ample opportunity that the reflected beams at some angle will also get captured. The techniques made the system very robust thus worked well with most kinds of rough surfaces and particularly proved suitable on the snow surface.

The transducer elements in this 3x3 receiver array are connected in series, as shown in figure 5b. This gives a very big advantage that the overall voltage received at the receiver section is a superimposition of the wavefront arriving at each of the receiver transducers. This strengthens the receiver signal, and even if only one transducer element has received the wavefront, it acts as a right input signal. The resultant signal is the sum of the entire signal received at individual piezoelectric crystals, giving better sensitivity on the receiver end. Using array ensures that the receiver has a better surface area and at least one of the receiver transducers shall receive enough amplitude of the returned echo to detect the surface.

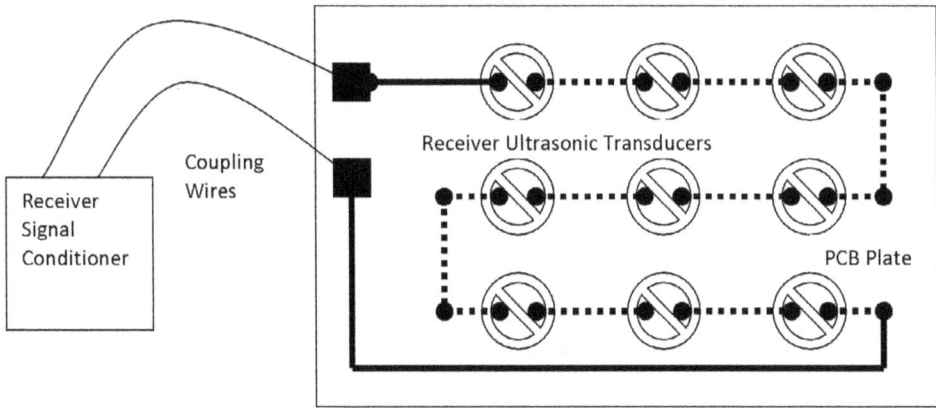

Fig [5b]: *Receiver transducer in the series array*

The receiver section, in this case, will receive the three echoes since there were three transmitted envelopes. This ensures that even if we are transmitting one envelop consisting of N number of pulses, the receiver wavefront will be a superimposed wavefront consisting of a 3xN number of pulses. This increases the system responsiveness and lessens the chances of missing the reflection. This will be clearer from the figure 6 which shows the three transmitted envelops received at the receiver.

Time shifted Superimposed Envelops

First Pulse
Crossing
Threshold

Threshold
Level

Zero Volts
Level

Noise
Superimposition

Echo from
Transmitter
Transducer 1

Echo from
Transmitter
Transducer 2

Echo from
Transmitter
Transducer 3

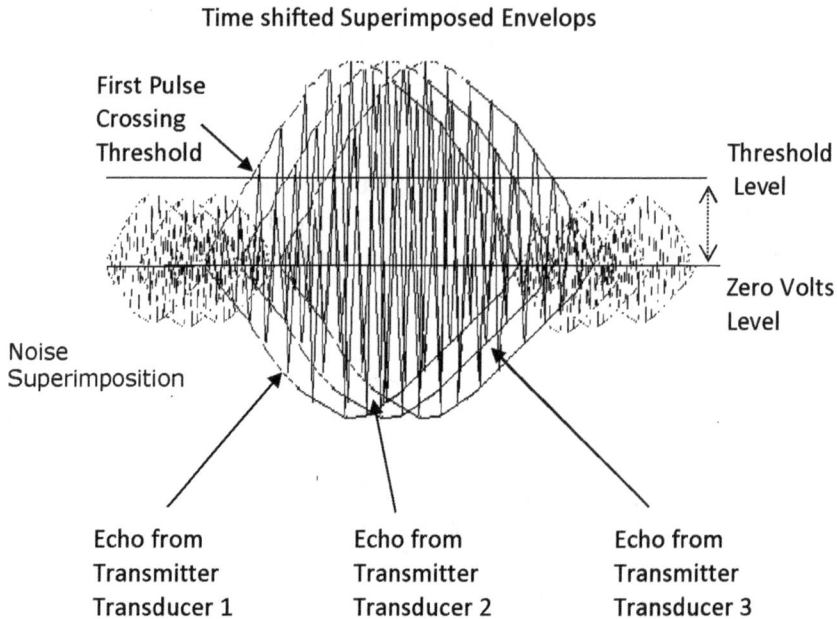

Fig [6]: *Superimposed time-shifted 3 transmitted wavefront envelopes*

5. THEORY BEHIND THE PHYSICAL IMPLEMENTATION OF TRANSDUCER ARRAY

The theory of array sources relies on mathematical derivations of aperture and frequency response. The treatment here is kept quite simple without including intricate derivation to relate it to the physical implementation of the ultrasonic array. Interested readers are advised to refer to Martin E. Anderson & Gregg E. Trahey's seminar [22] on ultrasound which covers array theory, 2-D Fourier transform, spatial frequencies, actions, and detailed computations.

Here we will touch the theory of transmitter and receiver linear array to highlight that the total response of the array is some mathematical function of all the individual transducers. Further, it will be highlighted how directional pattern, beam width, dispersion, several side-lobes, and combined array response depends upon the size, spacing, and geometry of the transducers. However, detailed derivations can be found in literature at reference [17, 20, 22].

To brief-up, the ultrasonic array is dealt in analogy with the optics. In optics, the Huygen-Fresnel principle states that wavefronts can be decomposed into a collection of point sources, each the origin of a spherical, expanding, a wave that can be represented as a free space Green's function. This concept underlies the derivation of an important tool, the Fraunhofer Approximation. The Fraunhofer approximation (FA) plays a pivotal role in our exploration of ultrasound *k*-space. Briefly, this well-known expression from the optics literature states that the far-field complex amplitude pattern produced by a complex aperture amplitude function is approximately equal to the 2-D Fourier transform of that function. Applied to ultrasound, this approximation states that the ultrasound beam's pressure amplitude pattern can be estimated by taking the 2-D Fourier transform of the transducer aperture. Naturally, this approximation is based on several assumptions and requirements that constrain its application to the far-field of an unfocused transducer or the focal plane of a focused transducer. As long as we do not violate the assumptions made in formulating the FA, this powerful approximation allows us to extend our intuition regarding linear systems to the study of ultrasound beamforming. Restricting the discussion for a moment to the lateral and axial dimensions, the most obvious example of an aperture function is the rectangular aperture of a linear array lying along the lateral coordinate axis, emitting a single frequency of sound [22].

5.1   *Frequency Response of Planar Transmitter Array*

Consider a planar transmitting piezoelectric array of an odd number (MxN) of identical elements, as shown in figure 7. In a general number of elements in X and Y directions are different, the i.e. array is usually made a rectangle [15, 16, 20, 21]. Also, the spacing in elements of X direction is different as the spacing of elements in Y direction. Also note that Δx and Δy is usually chosen to be less than the one-half wavelength to avoid grating lobes, that is, extraneous main lobes, under all condition of beam steering.

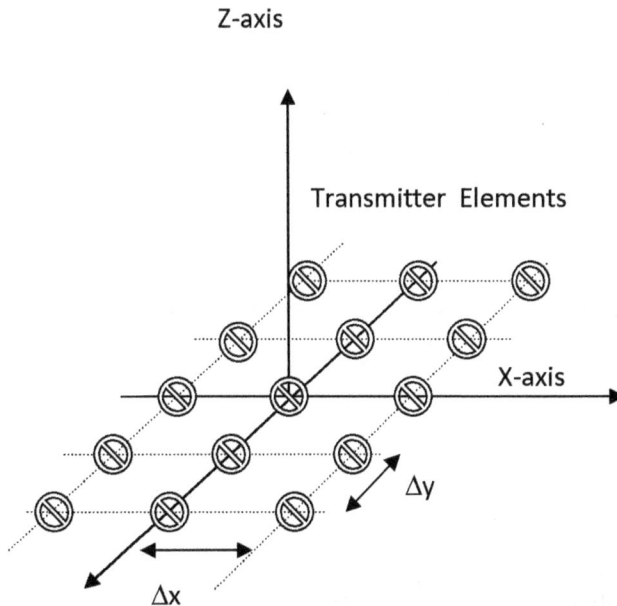

Fig [7]: Planar rectangular array arrangement of M x N (odd) transmitter elements

The frequency response of one ultrasonic element is a function of frequency 'f' and location (x, y). Let frequency response of one ultrasonic element located at (x, y) operating at a frequency 'f' is r (f, x, y). When several such elements are mounted in a plane, then the entire system has a resultant total frequency response indicated by A (f, x, y).

Using the principle of superimposition, we can conclude that the total frequency response of the array is the summation of the response of all the elements taken over all the locations. This is expressed as under:

$$A(f, x, y) = \sum_{j=0}^{M} \sum_{k=0}^{N} c_{jk}\, r(f, x - j\,\Delta x, y - k\,\Delta y) \qquad \text{--- eq (1)}$$

Where
j, k position count number of the element in x and y-direction w.r.t. to 0th element
M is the total number of elements in X direction
N is the total number of elements in Y direction
f is the frequency in Hz
Δx is the constant inter-element spacing in meters in X-axis
Δy is the constant inter-element spacing along Y-axis

And c_{jk} is the complex weight associated with the element (j, k). This complex weight of the individual responses or the coefficient of individual response by a particular element at the desired point is given by:

$$c_{jk} = a_{jk} \exp(+i.\theta)$$

Where a_{jk} is the real amplitude weighting term
              θ is the Real phase weighting term

The convolution theorem can be applied as under on term $r(f, x - j\,\Delta x, y - k\Delta y)$ to express it in the form of the shifted impulse response of point sources. If we take $\delta(x)$ as the impulse response of a point source/transducer element located at x and $\delta(y)$ as the impulse response of a point source/transducer element located at y, then:

$$r(f, x - j\,\Delta x, y - k\Delta y) = r(f, x, y) ** \delta(x - j\,\Delta x).\,\delta(y - k\,\Delta y)$$

where ** implies the convolution.

Putting this value in equation (1) we get

$$A(f, x, y) = \sum_{J=0}^{M} \sum_{k=0}^{N} c_{jk}\, r(f, x, y) ** \delta(x - j\,\Delta x) \cdot \delta(y - k\,\Delta y)$$

$$A(f, x, y) = s(x, y) ** r(f, x, y) \qquad\qquad\text{--- eq (2)}$$

where the term:

$$s(x, y) = \sum_{j=0}^{M} \sum_{k=0}^{N} c_{jk}\, \delta(x - j\,\Delta x) \cdot \delta(y - k\,\Delta y)$$

is defined as the frequency response of an equivalent planar array of M x N odd complex-weighted point sources. It contains all information containing the array, such as the total no of elements, the complex weight, geometry of array, positioning of the array in XY plane.

5.2 Far-Field Beam Pattern of Planar Transmitter Arrays

In transmit mode far-field beam pattern of this array is a measure of the ability of the array to concentrate the acoustic power in the preferred direction. When used in receive mode, the far-field beam pattern is a measure of the ability of the array to distinguish among several sources located at different spatial locations [14, 16, 20].

Let A (f, x, y) is the complex frequency response function (aperture function), as given by the equation (2).

From aperture theory, complex aperture function and the far-field beam pattern form a spatial Fourier transform pair is given as under [22]:

$$D(f, \theta, \psi) = F\{A(f, x, y)\} \qquad\qquad\text{--- eq (3)}$$

Where   D is the far-field directivity function or beam pattern, θ And ψ are the spherical angles as shown in figure 8.

From equation (2) and equation (3), we get

$$D(f, \theta, \psi) = F\{s(x, y) ** r(f, x, y)\}$$

Which implies

$$D(f, \theta, \psi) = S(\theta, \psi)\, R(f, \theta, \psi) \qquad\qquad\text{---eq (4)}$$

Because convolution in spatial domain gives products in spatial frequency domain. Here R (f, θ, ψ) is the double Fourier transform (DFT) of {r(f, x, y)} and S(θ, ψ) is the double Fourier transform of {s(x,y)}. In terms of spatial frequencies above equation can be written as under:

$$D (f, f_x, f_y) = S (f_x, f_y) R (f, f_x, f_y)$$ --- eq (5)

Where f_x and f_y are spatial frequencies in units of cycles per meter, given as

$$f_x = u / \lambda = (\sin\theta \cdot \cos\Psi)/\lambda$$
$$f_y = v / \lambda = (\sin\theta \cdot \sin\Psi)/\lambda$$

Here 'u' and 'v' are direction cosine with respect to X and Y-axis.

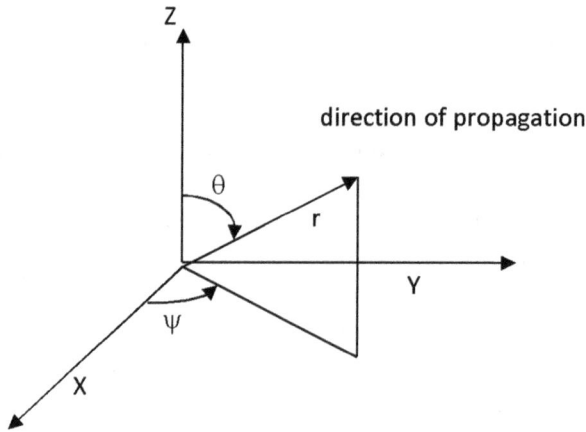

Fig [8]: *Spherical angles for far-field beam transmission*

Eq (5) is referred to as Product Theorem for Planar Arrays. *It states that the far-field directivity function of a planar array of identical elements is equal to the product of the far-field directivity function R (f, f_x, f_y) of one element and the far-field directivity function of an equivalent planar array of point sources.*

S (f_x, f_y) is in the form of Discrete Spatial Fourier Transform (DSFT), i.e. transform in the form of sines and cosines, also called Wannier transform, of a complex number. The real part of amplitude weight c_{jk} controls the shape of the far-field beam pattern, i.e. the width of the main lobe and the level of the side-lobes. The phase weight θ allows the steering or the tilting of the beam in the preferred direction.

5.3   *Frequency Response of Rectangular Receiver Array*

Considering a case of a planar array of M x N (odd) identical, complex-weighted point sources lying in the X-Y plane as shown in figure 9. The array is being used in the receiver mode. Based on the output electrical signal from the individual element in an array, it is desired to estimate both the target's direction and the frequency contents of the radiated acoustic field [17, 21].

Assume that output electrical signal from element (m, n) of an array is y(t, r) which can be written as y(t, m Δx, n Δy) also since element (m, n) is located at x = m Δx , y = n Δy as shown in figure 9. Here m = -(M-1)/2, ...0,....+(M-1)/2 and n = -(N-1)/2,....0.....+(N-1)/2 for M and N odd.

Adding complex weight to above output, the output from one element becomes:

$$Y(t, x, y) = C_{mn} \, Y(t, m \, \Delta \, x, n \, \Delta y)$$

Outputs from all the elements are

$$Y(t, x, y) = \Sigma \Sigma \, C_{mn} \, Y(t, m \, \Delta \, x, n \, \Delta y) \qquad \text{--- eq (6)}$$

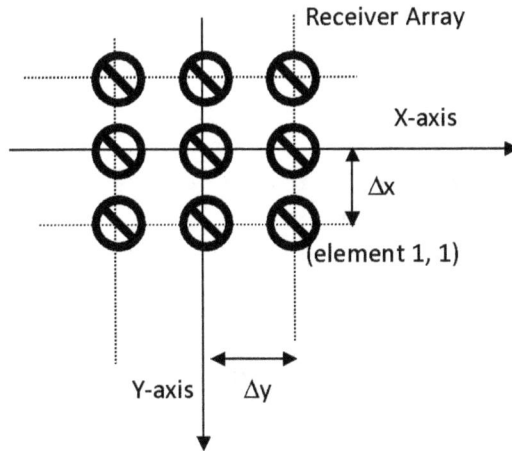

Fig [9]: *Rectangular receiver array in 3 x 3 configuration*

A treatment on this equation similar to eq (1) will cause frequency and angular spectrum is given by:

$$Y(\eta, \gamma_x, \gamma_y) = F_t \, F_x \, F_y \{y \, (t, x, y)\} \qquad \text{--- eq (7)}$$

This equation again shall give rise to equations of DSFT similar to as derived for the case of the transmitter. From the detailed computations of arrays, *it can be concluded from the plot of the above frequency response equations that array has a region of maximum response to signals centered at θ=0 (broadside) This region of maximum response is usually called main lobe and the other maxima is called side-lobes. The response drops off rapidly as the signal is moved away from broadside* [22, 23, 24].

In the resultant response of the entire array, to avoid grating lobes (extraneous main lobes) under all conditions of beam steering, inter-element spacing Δx and Δy must satisfy Nyquist condition:

$$\Delta x \text{ and } \Delta y < \lambda_{min} / 2 \qquad \text{--- eq (8)}$$

6. IMPLEMENTATION OF TRANSDUCER ARRAY

In the present case, design comprises a planar array of 3 x 1 transmitter array and 3 x 3-receiver arrays of ultrasonic transducers. The characteristic frequency is 40 KHz with a wavelength of 9-11 mm (varies with the velocity of sound), and diameter of the element used is 12 mm. Inter-element spacing is calculated from equation (8) is maximum 5 mm. However, the inter-element spacing of 5 mm is impossible when the diameter of the transducers is 13 mm. The diameter of the transducers cannot be selected too less because the directional pattern depends upon D/λ ratio. In the current case, the D/λ ratio of 1.2 is associated with each of the transmitting element.

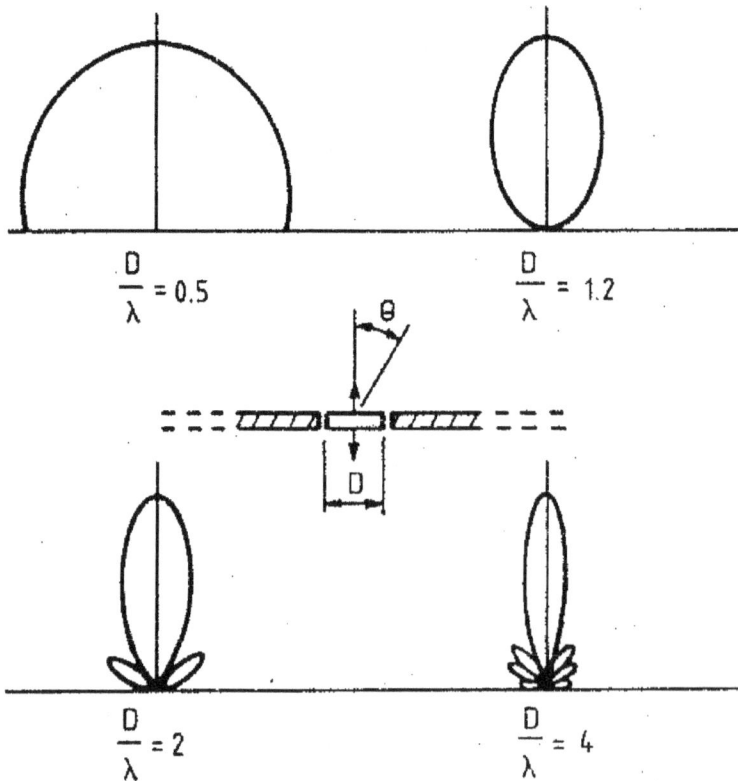

Fig [10]: *Effect of diameter to wavelength ratio of the radiator on the directional pattern of the ultrasonic beam* [14, 15, 16]

It can be seen from the figure 10 how D/λ governs the beam pattern of an individual element. Too less value of D implies multiple-grating lobes generated by each element which are not acceptable in this case.

So, the only way is to adjust the center to center inter-element spacing at an optimized value. We fixed it approximately at λ. A nominal value of 13 mm inter-element spacing is good enough to a tradeoff between combined resultant multiple-grating of the complete array and gratings of the individual transducer element. Incidentally, this is the minimum spacing allowed due to the size of the transducers. However, our experiments indicate that the number of gratings increases as we increase the spacing and hence the receiver gets triggered with the wrong reflection and responds to the grating beam itself.

The resultant directional characteristics for the unidirectional response of the combined array are shown in figure 11 in the form of the plot of acoustic pressure vs. angle vs distance at all the points around the radiators [15, 20]. The directivity is surely obtained, but the transmitting beams utilize a larger surface area at the surface where it strikes and hence the probability of considerable reflection even at a large range is increased proportionately. Figure 11 shows the unidirectional response of a transmitter array at 40 KHz having element diameter D= 1.2λ.

Directional pattern of this array consists of two zones, namely near zone and far zone. The side-lobes in the near-field zone do not get completely removed from the directional pattern as it should have been as ideally shown in figure 10 for D/λ =1.2.

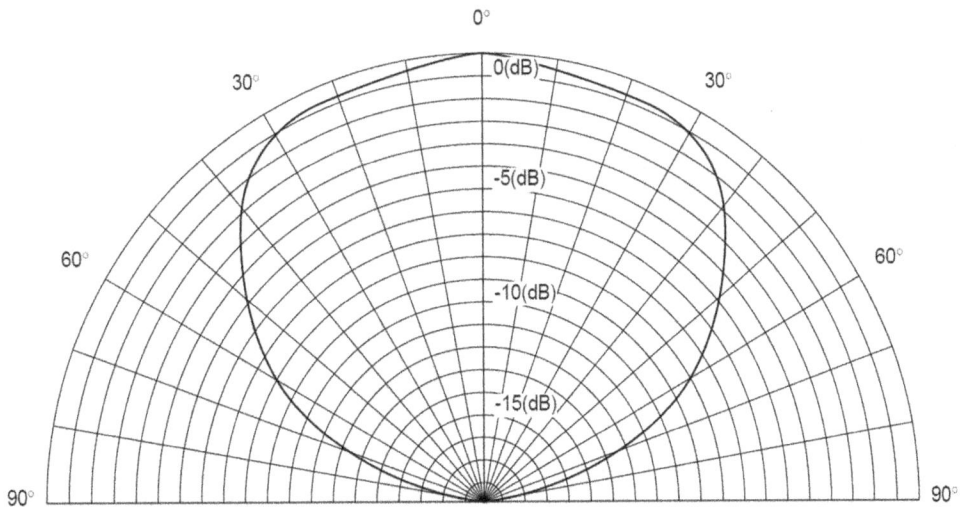

Fig [11]: *Improved directional response using array transducers with D/λ =1.2, $\Delta x=\Delta y= \lambda$ (D is diameter of one element and λ is the wavelength, Δx & Δy inter-element spacing)* [15, 17, 18]

The number of maxima is D/λ, currently a number greater than 1. The D/λ ratio not being exactly equal to 1.2 and also since the inter-element spacing does not satisfy the

Nyquist condition, two maxima (side-lobes) around the main maxima are observed. In case of odd number, a maximum is obtained in the center with double the mean acoustic pressure. The acoustic pressure along the axis of the oscillator fluctuates between zero and double means value.

To remove these side-lobes, there could have been following three solutions:

a)   Use of lower frequencies could have resulted in no side-lobes, but that would have given non-directive response [8]. Optimum frequency of 40 KHz has been used.

b)   Proper shielding of the transmitting transducer and receiving traducers so that near filed do not trigger the receiver. Too long shielding would create severe interference at the sensor hood itself.

c)   The effects of near-field side-lobes could be removed through electronics design and software implementation by not allowing the receiver to "see" anything in the near-field. This could have impaired the near-field detection.

In our application, we used all the three solutions with proper tradeoffs. The 40 KHz frequency has been optimally chosen. The reasons for choosing this frequency are many. Fist reason being the availability of transducers in 40 KHz frequency range with broader directional pattern response. At 40 KHz ultrasonic have relatively less absorption in the material surface and range is nominally good at around 4-5 meters. Further using 40 KHz transducers do not sacrifice the accuracy.

In the case of transmitting-receiving operation, the transmitter and the receiver are fitted with coupling adapter and two probes are carefully shielded from each other both electrically and acoustically. Mechanically both probes have been combined in a single unit connected to the instruments by a twin cable.

The far-field is simpler as compared to the near-field. The steep maximum at the end of the near-field widens with increasing distance. The figure 12 shows that the pattern opens at a definite angle, which is obtained for the first zero points, by connecting it to the center of the radiator along the broken lines. The angle of one of these lines with the axis of the radiator is the angle of divergence or beam dispersion. The angle of divergence is γ_0 given by the theory of diffraction:

$$\text{Sin } \gamma_0 = 1.2 \, \lambda \, / \, D$$

Above formula is valid only for the small value of λ/D i.e. only small values of angles of divergence is obtained correctly.

The figure 12 shows the beam dispersion [20]. The above equation gives approximately a dispersion of around 60° with D= 12 mm and λ = 9 mm, which is double as compared to normal applications. This further increases the area of front exposure. The power attenuation due to dispersion is compensated through greater voltage amplitude at the transmitter for generation of ultrasonic pulses.

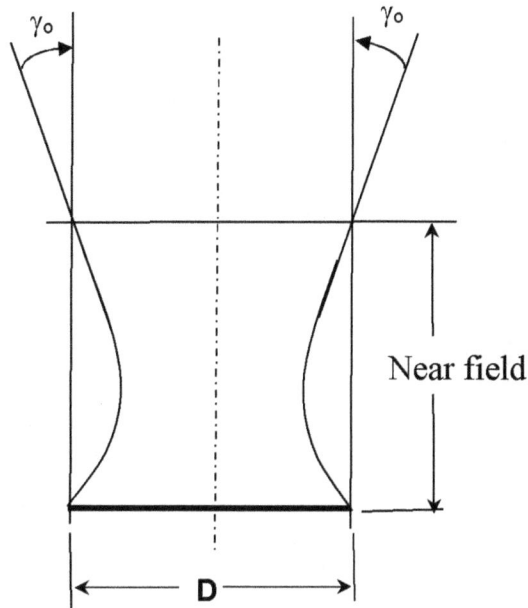

Fig [12]: *Beam divergence in the far-field with D/λ= 1.2*

7. CONSIDERATIONS FOR IMPROVEMENT OF DIRECTIONAL RESPONSE

Directional characteristics of the array include the angle of divergence and beam width of the emitted signal. The measure of the directional pattern is beamwidth and directivity factor or directivity index. When applying the echo method, the sensitivity of array is also very important than the directional characteristic of the radiated sound field. The sensitivity characteristics in the echo method are equal to the square of the directional characteristics of the sound field. Directional response and sensitivity of the transducer array are greatly affected by geometry, shape, diameter, and area of the radiator. The magnitude of the acoustic pressure at a given distance is determined by the ratio of area to the wavelength of the radiator. This, in turn, creates dependency on the diameter of the element. Shape too has great effects on the directional pattern. The conical surface radiator may have many side-lobes while the rectangular radiator is no longer axially symmetrical. It becomes broad in the plane, which contains the axis and narrow side of the rectangle, and vice versa. Circular disc radiator has more flattened and freer from the directional pattern as shown in the figures. At a great distance, sound field follows the distance law of spherical wave, i.e. the acoustic pressure decreases inversely with the distance. Out of this at the design stage, we cannot control any of the factor except wavelength, i.e. frequency of the transducer. The diameter of the transducer is prefixed, however; the transducer with the right diameter could be selected with the right frequency.

The governing factor is diameter to wavelength ratio. As seen above the ratio of oscillator diameter D to the wavelength λ determines the spread of the interference field and several maxima and minima. The character of acoustic beam is determined by the ratio of D to λ. By making this value large, we get a sharply defined and far-extending beam, but several gratings also increase. Beam becomes narrow if the diameter of the radiator is increased. With small diameters, the angle of divergence increases for the same wavelength. Further, if higher frequencies of the order of 100 kHz (smaller wavelength) are used, the narrower beam pattern is obtained.

For a radiating array, there can be a dramatic effect in response by varying the amplifier gains for transmitting voltages. Since three set of transducers in transmitter array are used, pulses three times in number as compared to transmitted pulses shall be received with the different amount of phase shifting [17,19]. The first pulse crossing the threshold gives a fair amount of accuracy of distance measurements. So, the random phase shifts in the received signal is not a problem.

Occasionally focusing probes of special design are used to increase the sensitivity over a definite range [21]. For this purpose, either a curved, ground, piezoelectric plate off ceramic material is used, or a curved layer with lens effect is cemented to the flat crystal plate [24]. The latter method greatly increases the sensitivity.

8. CONCLUSION

The experimental results have been very favorable. With 40KHz ultrasonic transducers arranged in array fashion, a considerable amount of reflected signal is received through arrayed receivers, which can be detected with the help of the suitable electronics design. The surface distance over 5 meters can be fairly detected despite the uneven, rough, porous snow surface which normally does gives problems of scattering and absorption of energy into the surface. However, the range depends purely upon the independent transducer specifications. If the transducers are combined in a suitable array, the results are more encouraging and reliable. The directional patterns can be made broader to offset the effects of non-smoothness of the surface. However, the suitable tradeoff with ultrasonic power per square centimeters should also be considered before shaping the beam width pattern. This non-contact remote measurement of level or depth can be made very effective with above design modifications incorporated in it.

REFERENCES

[1]   Gooberman G.L., Pulse Techniques, Ultrasonic Techniques in Biology and medicine, Illiffe books Ltd, London, 1967.

[2]   Satish Kumar et al, Snow Depth Senor, Proceeding of National symposium on Sensors and Transducers, 1996.

[3]   Mellor, M., Engineering properties of snow, Journal of Glaciology, volume 19, pg 15-66, 1977.

[4] Yamazaki, Kondo, T. J., Sakuraoka,T. and Nakamura, T., A one dimensional model of evolution of snow cover characteristics, Journal of Glaciology, Vol. 18, pg 22-26,1993,

[5] Hall, Derothy K., Remote sensing of Ice and Snow, London, Chapman and Hall Publications, 1985,

[6] Combs, Charles M.; Goodwin, Jr., Perry H., Adjustable ultrasonic level measurement device, United States Patent 4221004, Aug 1978

[7] M. Krause, et. Al. Comparison of Pulse-Echo-Methods for Testing Concrete, E-Journal of Non-destructive testing, Vol.1 No.10, October 1996,

[8] A. Hämäläinen and D. MacIsaac, Using Ultrasonic Sonar Rangers: Some Practical Problems and How To Overcome Them, Phys. Teach. Vol 40, pp 39, 2002.

[9] Ilene Busch-Vishniac, Elmer Hixson, Acoustical Instrumentation, Encyclopedia of Applied Physics vol 1, VCH publishers, , pg 63-88, 1991

[10] Heizfield K.K & Litovitz T.A Absorption & dispersion of Ultrasonic wave, 1959.

[11] Mason W.P, Properties of Gas, Liquids and Solids", Physical Acoustics, vol 2, 1965

[12] Balantine D.S et al, Acoustic Wave Sensors-theory, Design and physico-chemical applications, Academic Press, 1997

[13] Ensminger D, Ultrasonic – the low and high intensity applications, Marcel Decker Inc, New York, 1973

[14] Busch I, Huxson E, Ultrasonic, Encyclopedia of applied Physics, VCH Publishers, vol 1, pp 63-88, 1991

[15] Ultrasonic, Encyclopedia of Physical Science and Technology, Vol 12, pp 662-664, Mcgraw Hill, 1982.

[16] Papadakis E.P., Physical acoustic principles and methods (W.P. Mason , ed.), vol 4 B, Academic, NY, 1968

[17] Hudson J.E, Adaptive Array Principles, Peter Peregrinus , London 1981

[18] Piezoelectric Ceramic Sensor (Piezoliote), Cat-P19-E8, Murata Manufacturing Co. Ltd, Japan, http://www.murata.com/catalog/p19.pdf

[19] Harold Carey, Reducing Side lobes of SRF10, Robot Electronics, Inc. http://www.robot-electronics.co.uk/htm/reducing_sidelobes_of_srf10.htm.

[20] John Szilard, "Ultrasonic", Encyclopedia of Physical Science and Technology", vol 14, Academic, London, (1987), pg 191-209,

[21] Edmund J.Sullivan, Acoustic Signal Processing, Encyclopedia of Physical Science and Technology, Vlo1, Academic Press- Orlando, Florida, 1987.

[22] Martin E. Anderson and Gregg E. Trahey, A seminar on k-space applied to medical ultrasound, Duke University, April 12, 2000, http://dukemil.egr.duke.edu/Ultrasound/k-space/bme265.htm.

[23] Ziomek L.J., Underwater Acoustic, A linear systems theory Approach, Academic Press- Orlando, Florida, 1985.

[24] Lawrence J. Ziomek, Underwater Acoustic, Encyclopedia of Physical Sciences and Technology, Vol 1, Academic Press- Orlando, pg 183-190, 1987.

[25] Attri, RK 2018/2005, 'Design of a Reliable Remote Surface Detector Based on Ultrasonic Pulse- Transit Technique to Detect Uneven & Non-Smooth Porous Snow Surfaces,' R.Attri Instrumentation Design Series (Snow Hydrology), Paper No. 1, *Research and Design of Snow Hydrology Sensors and Instrumentation*, 2nd edn., pp. 1-23, Speed To Proficiency Research: S2Pro©, Singapore.

[26] Attri, RK 2018/1999, 'Design strategy of snow depth sensor based on ultrasonic pulse-transit technique for remote measurement of snow cover thickness,' R.Attri Instrumentation Design Series (Snow Hydrology), Paper No. 2, *Research and Design of Snow Hydrology Sensors and Instrumentation*, 2nd edn., pp. 25-41, Speed To Proficiency Research: S2Pro©, Singapore.

FURTHER READINGS

- Szilard J, Ultrasonic testing – non-conventional testing techniques, Willey, NY, 1982
- Silk M.G., Ultrasonic transducers for non-destructive testing, Adam hilger Ltd, Bristol, 1984
- Hueter and Bolt, *Sonics*, Wiley, NY, 1955
- Landee, R.W. et al, *Electronics Designer's handbook*, Mcgraw Hill , 1957
- P. Boltryk , M. Hill , A. Keary , B. Phillips , H. Robinson and P. White, An ultrasonic transducer array for velocity measurement in underwater vehicles, Ultrasonics, Volume 42, Issues 1-9, April 2004, Pages 473-

478

- M. G. Maginness, J. D. Plummer, W. L. Beaver, and J. D. Meindl, *State-of-the-art in two-dimensional ultrasonic transducer array technology,* Medical Physics, Volume 3, Issue 5, pp. 312-318, September 1976
- Walter Patrick Kelly, Jr. , Rodney J Solomon, Two-dimensional ultrasound phased array transducer, United States Patent 6894425, May 17, 2005, http://www.patentstorm.us/inventors/Walter_Patrick_Kelly,_Jr_-1324371.html.
- S. Smith et al, "2-D Array Transducers for Medical Ultrasound at Duke University: 1966", ISAF '96 Proceedings of the 10th IEEE Int'l. Symposium on Appl. of Ferroelectrics, vol. 1, Aug. 1996, pp. 5-11.
- T. Miyashita, T. Itaya and T. Matsumoto, "Reconstruction of Wide-Bandwidth Scattering Responses from Narrow-Bandwidth Ultrasound Echo in Air," Japanese Journal of Applied Physics, vol.38, pp.3135-3138 (1999).
- 10W. J. Hughes, W. Thompson, Jr., and R. D. Ingram, "Transducer array scanning system," United States Patent 3905009, Sept. 1975.
- Susan Dumbacher et.al, Source Identification Using Acoustic Array Techniques, Proceedings of the SAE Noise and Vibration Conference, *Vol 2, pp 1023-1035, Traverse City, MI, May 1995*
- Charles W. Danforth, *Acoustic Applications of Phased Array Technology,* 1998, http://casa.colorado.edu/~danforth/science/sonar/sonar1.html.

Paper No.4

SNOWPACK TEMPERATURE PROFILE SENSOR

RAMAN K. ATTRI

EX-SCIENTIST,
CENTRAL SCIENTIFIC INSTRUMENTS ORGANIZATION INDIA

The previous version of this paper was presented and published as:
Shamshi, MA, Attri, RK & Sharma, VP 1996, "Snow pack temperature sensor," *Proceedings of National Conference on Sensors and Transducers*, Chandigarh, pp. 180-189, viewed 24 Jan 2018, <https://www.researchgate.net/publication/275276742>.

Abstract - *The study of snow hydrological characteristics plays the major part in a forecast of snow-melt, snow run-off water of rivers, the potential release of snow avalanche and climatic changes. The increasing importance of the study of snow hydrology has led to the development of many snow sensors which monitors a physical parameter of snow. Design of such snow sensors has been a difficult job because of harsh environmental conditions. This paper presents the design considerations and performance characteristics of snowpack temperature profile sensing probe. This probe provides the temperature gradient and temperature distributions within the snowpack and ground at different depths. The information on temperature distribution in snowpack is very important for conducting exchange studies of snow. These studies serve as a tool for snow avalanche forecasts.*

1. INTRODUCTION

Snow is defined as falling or deposited ice-particles formed mainly by sublimation. It may be considered as solid precipitation composed of ice crystals, falling in the air or deposited on the ground.

Snow cover builds up layer by layer from one storm to another through interactions with meteorological parameters. The interactions of meteorological parameters mainly relate to the energy exchange between snowpack and environment, between snowpack and ground, and within the snowpack. This energy exchange alters the thermal regime and hence, the morphological state which governs the mechanical properties and therefore, the stability of a snowpack. Thermal properties of snow are used to characterize the varying conditions of the snowpack. It governs the quantity of water storage in the basin, snow run-off water, evapo-trans-piration rates, wet snow avalanche, etc. The conclusions arrived at through these hydrological characteristics are used in the analysis, modeling of water resources, weather forecast and change in ambient temperature, etc. Therefore, the study of energy distribution in a snowpack requires a lot of attention. The energy distribution within snowpack, snow-air interface, the snow-ground interface is derived from the data of snow temperature at different points i.e. from the temperature gradient and temperature distribution.

This paper describes the design of a snowpack temperature profile sensing probe along with microcontroller-based data-logger developed to study the temperature distributions within the snowpack. From the data received, the temperature gradient within the snow cover, ground and air are calculated.

2. SIGNIFICANCE OF DATA RELATED WITH SNOW PACK TEMPERATURE DISTRIBUTION

The energy distributions concluded from the temperature distribution data is used -
- To evaluate the snow, melt based on the snow-air energy exchange.
- To develop snow-melt run-off models of river based on energy exchange approach.
- To study the possibility of melt and destruction of the bond between ground and snowpack based on energy distribution within snowpack & snow-ground energy exchange.
- To study the process of metamorphism of snow for evaluating the strength and stability of snowpack.
- To understand the formation and release of an avalanche.
- To forecast the weather changes due to changes in snow cover and snow melting.

Snow being a thermodynamically unstable material undergoes morphological changes during the development of different forms of crystals within the snowpack. The shape and size of the bonding, packing, etc. control the mechanical properties and hence stability and strength of snow cover thus evolved. The rate at which this process of metamorphism proceeds depends upon average temperature, temperature gradient, and porosity, etc. The energy exchange at ground interface determines the bond between ground and snow. This gives a potential guess of snow avalanche release. The density and hence, snowbound mass are also related to crystal growth which in turn depends upon temperature gradient of different layers.

Snow-melt produces some free water, the movement of this water vertically downward increases the rate of metamorphism, reduces mechanical strength and deformation resistance of snowpack by weakening inter-granular bonds. With the increase in free water content, the snow tends to lose cohesion, and inter-granular bonds start to disintegrate. This causes a drastic decrease in shear strength. The study and forecasting of snow-melt water are necessary to forecast wet snow avalanche. For this purpose, temperature distribution study is essential.

Further, development of snow-melt run-off models of rivers is also based on energy exchange approach. This no doubt predicts the river water level. It is the most important conclusion in the case of irrigation and flood risk assessment.

3. SNOW TEMPERATURE SENSOR SYSTEM DESIGN REQUIREMENTS

This paper aims to explain the design and development of a probe which measures the temperature of snowpack at different points along with the ground temperature and air temperature accurately.

A suitable electronics and mechanical design have been worked out to meet the requirements in terms of accuracy, performance and environmental conditions to achieve the best possible performance and stability.

The snow profile temperature sensor system is to be installed in heavily snowbound areas. So, the design requirements have been severe. The most important consideration in the overall design and mechanical construction of the system has been the environmental conditions. The probe has to operate round the clock for several months under these conditions: -

- Temperature range of operating is -50°C to + 50°C.
- Relative humidity to be handled at the place of measurement is likely to be 100%.
- Wind speed of the order of 200 Km/h is to be handled.
- Moisture ingressing in the system should be nil.
- The system should be unaffected by rain or snowfall.
- High rigidity, strength and tight packaging of the system.
- Accuracy in temperature measurement should be 0.1 deg C.
- Resolution of temperature sensing probes is to be 0. 1 deg C.
- Current consumption to be less than 10 mA.

Because of these severe environmental conditions, the packaging should be water and moisture proof & wires and cables are to be selected so that these can withstand these operational conditions.

The current consumption has to be very low because the system is field operated on a battery.

4. SYSTEM DESCRIPTION

Proposed snowpack temperature profiler consists of a total of 28 temperature sensors, 20 of which are snow sensors measuring temperatures of different layers of a snow pack as it builds up, 7 of them are ground sensors measuring ground temperatures at various depths in the ground and 1 is air temperature sensor. Platinum Resistance Temperature Detector (RTD) has been selected to be used as sensing elements. Each sensor is integrated with an analog electronics unit as shown in figure 1(a). This electronics unit contains the signal conditioning circuit for each sensor which converts the sensor resistance into an equivalent voltage signal. The linearization and calibration of RTD have been done in the analog unit. The PCBs for component mounting are gold plated to make it resistant to water/moisture. High precision, high-performance military-standard components have been used in these units.

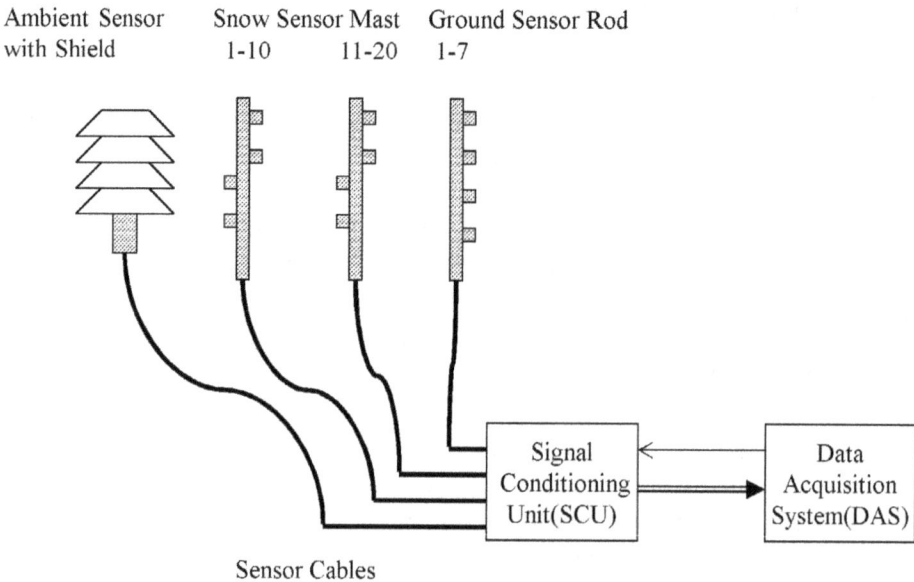

Fig [1a]: 28 temperature sensors interfaced with signal conditioning unit and data acquisition system

Each sensor has its own analog unit referred to as a channel. All the 28 channels are then interfaced to a data-logger, see figure 1(b). The data-logger is 35 channel data storage modules. It is a microcontroller-based unit which controls the sampling interval and multiplexing of 28 channels. The multiplexed signal is given to A/D converter which converts each analog value to the digital format. This digital data corresponds to temperature read by the corresponding sensor. This value is suitably computed by the microcontroller and converted into an engineering unit. The value is stored in memory locations along with sensor identification number and time of recording. The reading of temperature sensor is taken after selectable timing interval. After every hour, all the samples are computed to find out MINIMA, MAXIMA & AVERAGE temperature of each sensor. The computed data is important and is stored in a separate plug-in/plug-out memory module. Data can be retrieved when required from these modules by reading them in memory reader in the laboratory. Provision of data retrieval from a data-logger through a computer interface is also given.

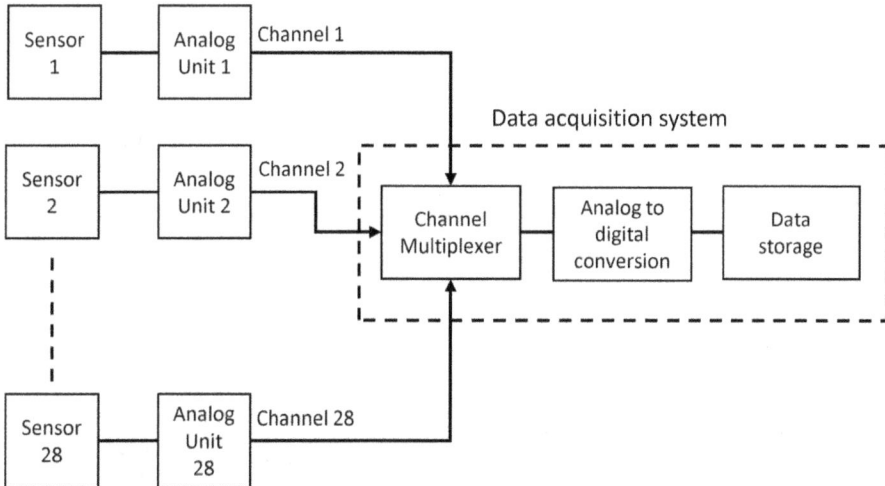

Fig [1b]: Block diagram of the snowpack profile temperature sensor

Application software has been developed which plots the temperature vs. height of the sensor at a different time as shown in figure 2. This gives air temperature, snow temperature profile as well as ground temperature variations with time. This supports energy exchange computation & hydrological inferences.

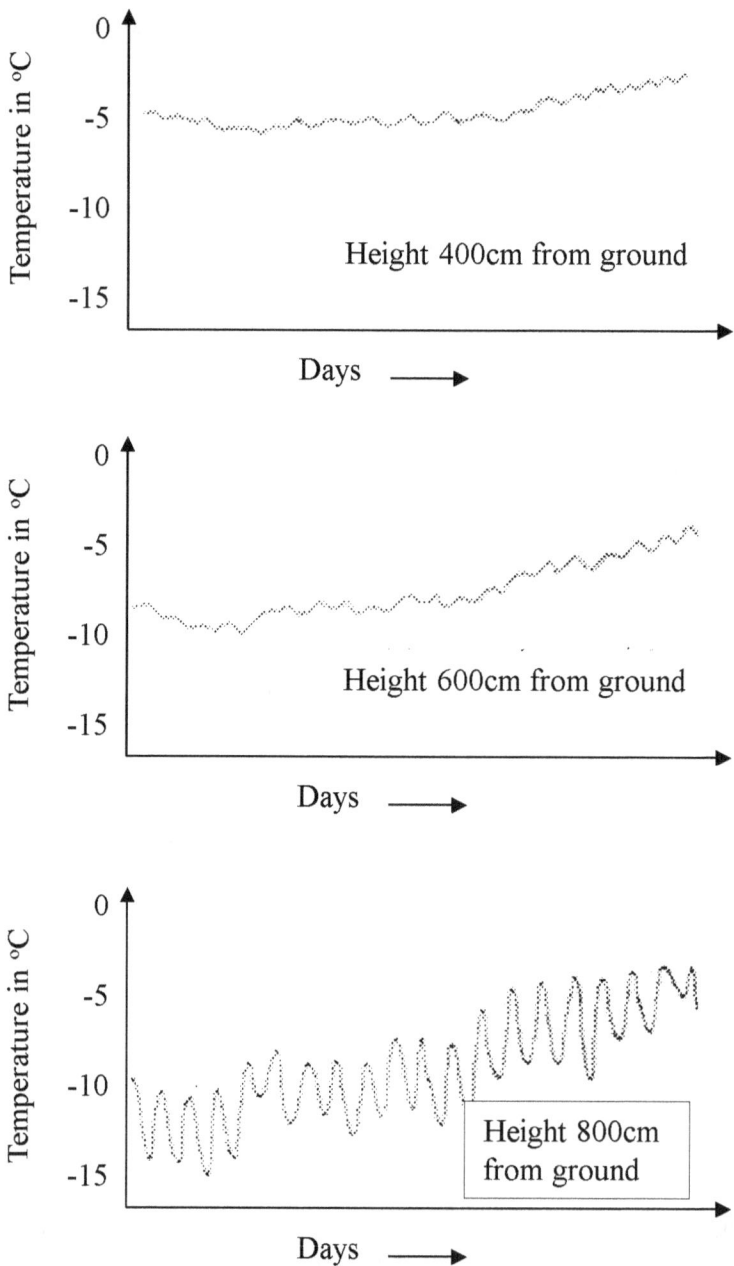

Fig [2]: *Variations of snowpack temperature with time at different heights from the ground*

5. *MECHANICAL DESIGN CONSIDERATIONS*

Proposed snow temperature profile sensor consists of a total of 28 temperature sensors. Out of these sensors, there are 20 snow temperature sensors, 7 ground sensors, and 1 ambient temperature sensor. The installation arrangement of 28 sensors has been shown in figure 3a. A total of 28 sensors have been mounted on a PVC tube of total 5m length. 1m of this PVC is buried inside the ground. The seven sensing probes with tips coming outside the holes are mounted tightly on this PVC tubes at heights -5.0, -10.0, -20.0,30.0, -50.0 and -100.0 cm relative to the ground. These 7 ground sensors provide the temperature profile of the ground and the temperature gradient at the ground interface.

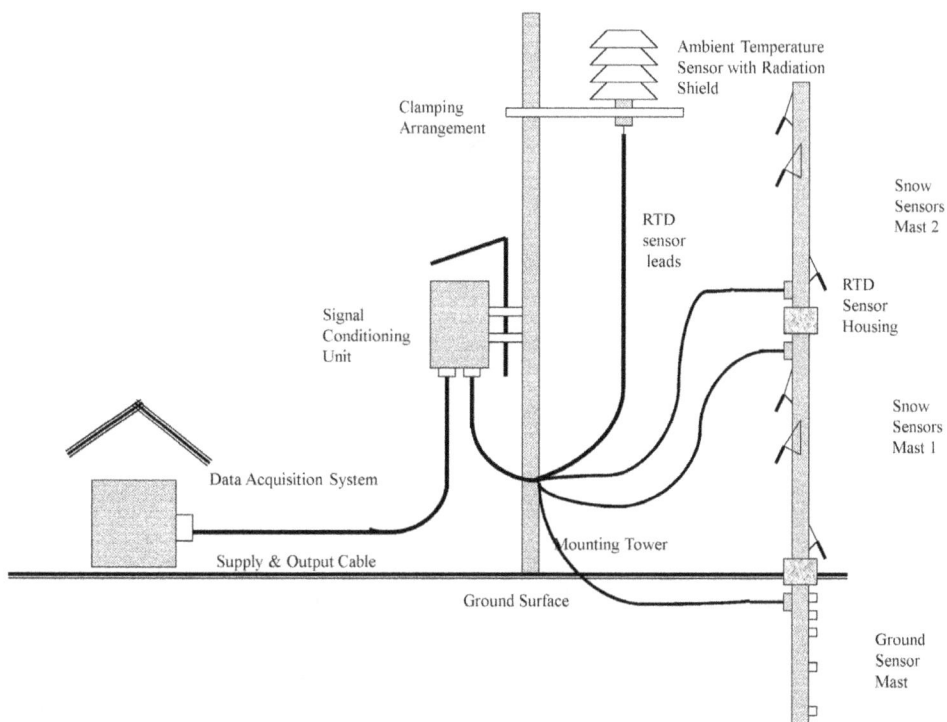

Fig [3-a]: *Complete mechanical arrangement of the snow temperature profile sensor system*

20 snow sensors have been mounted on the remaining 4m of PVC mast outside the ground at the spacing of 20 cm each. However, the mounting arrangement of snow sensor probes is entirely different from the mounting arrangement of the ground sensor as shown in figure 3b. Snow sensor probes are 89 mm in length & 4.75 mm in diameter. They are fixed on a PVC assembly of triangular shape so that the sensor makes 45° slanting downward angles with PVC mast. Further, adjacent sensors are 300 apart in the

horizontal plane. This kind of arrangement has been designed because of the following three reasons –

i.   450 slanting angles of the sensor so that snow may slide down & do not remain deposited over the probes during snow depletion.

ii.   300 horizontal spatial distance to cover all directions around PVC mast for temperature measurements.

iii. 3-dimensional coverage of snow for temperature measurements.

The extension leads from all sensors are restricted in thickness and resistance because PVC is 5m long and the diameter is just 3 cm. It limits the number of extension wires that can be passed through the tube. The resistance of extension leads affects the accuracy of the system.

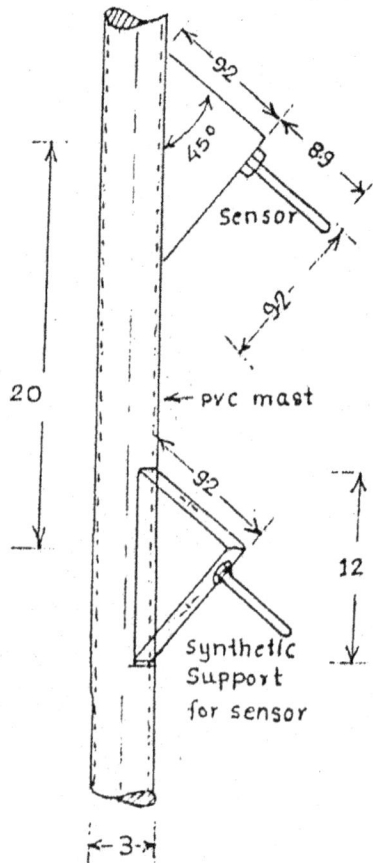

Fig [3-b]: Enlarged view of snow- sensor PVC mast'

The sensors and PVC tube are painted shiny white to reflect all solar radiations. One sensor is used to monitor the ambient temperature above the snow cover. This sensor is covered with a self-aspirated radiation shield so that sensor gives the true air temperature and is not affected by solar radiations. This sensor is either to be mounted on the top of PVC tube or the supporting tower.

The ambient temperature sensor has the flexibility of increasing or decreasing the height of the sensor as the snow cover evolves or depletes off.

Leads from all the sensors go to a junction box where the analog unit of the system is mounted. The junction box is mounted at such a height that it always remains out of now cover. The analog output is passed through low-temperature cables to a 35-channel data-logger, where digital data is stored.

The overall mechanical design has been made tight so that system handles specified temperature range, enormous wind speed and humidity up to saturation. Military-standard moisture proof connectors, components, assemblies, and leads have been used. Special efforts have been made not to make the system bulky. Transportability has also been given due consideration.

6. PERFORMANCE AND EVALUATION

The sensing part controls the overall performance, accuracy, and linearity of any instrumentation system. Therefore, one of the most critical considerations in the electronics design of the temperature probe is the selection of the sensing element. Accuracy, linearization, performance & stability of the temperature measuring system depends primarily upon the sensor performance and remedial measures taken against any error arising. Because of the harshest environment encountered, the performance of all available temperature sensors has been critically analyzed. After studying their relative performances, platinum RTD has been selected to be used for sensing.

The self-heating error, extension lead error, non-linearity error, and reference voltage instability error have been eliminated by incorporating suitable electronics design. The critical system performance is under testing & analysis.

7. CONCLUSION

Snow profile temperature sensor system has been designed and tested considering the accuracy and reliability of data obtained from there. The conclusions fetched from temperature distribution study has been employed in hydrological studies of snow and forecast of snow manifestation. The mathematical models for above hydrological study and forecast are complicated functions of other parameters also viz. snow depth, snow density, snow surface temperature, water equivalent, albedo, thermal conductivity, porosity, relative humidity, and wind speed, etc. So, a complete weather station employing wind speed sensor, relative humidity sensor, snow depth sensor, snow surface sensor, etc. has to be an integrated snow profile temperature sensor for a complete study of snow. These sensors are also under development leading to a stipulated weather station for snow hydrologists.

REFERENCES

[1] Attri, RK 2018/1999, 'Design of an Instrumentation System to Record Distribution Profile of Snow Layer Temperature for Modelling of Snow Avalanche Forecast,' R.Attri Instrumentation Design Series (Snow Hydrology), Paper No. 5, Research and Design of Snow Hydrology Sensors and Instrumentation, 2nd edn., pp. 66-75, Speed To Proficiency Research: S2Pro©, Singapore.

[2] Attri, RK 2018/2000, 'Design Approach to Use Platinum RTD Sensor in Snow Temperature Measurements,' R.Attri Instrumentation Design Series (Snow Hydrology), Paper No. 6, Research and Design of Snow Hydrology Sensors and Instrumentation, 2nd edn., pp. 66-75, Speed To Proficiency Research: S2Pro©, Singapore.

[3] Attri, RK 2018/2000, 'Practical Design Considerations for Signal Conditioning Unit Interfaced with Multi-point Snow Temperature Recording System,' R.Attri Instrumentation Design Series (Snow Hydrology), Paper No. 7, Research and Design of Snow Hydrology Sensors and Instrumentation, 2nd edn., pp. 66-75, Speed To Proficiency Research: S2Pro©, Singapore.

[4] Attri, RK 2018/2000, 'Design of A True Snow Air Temperature Sensing Probe,' R.Attri Instrumentation Design Series (Snow Hydrology), Paper No. 8, Research and Design of Snow Hydrology Sensors and Instrumentation, 2nd edn., pp. 66-75, Speed To Proficiency Research: S2Pro©, Singapore.

[5] Shamshi, MA, Attri, RK & Sharma, VP 1996, "Snow pack temperature sensor," Proceedings of National Conference on Sensors and Transducers, Chandigarh, pp. 180-189, viewed 24 Jan 2018, <https://www.researchgate.net/publication/275276742>.

[6] Attri, RK, Sharma, BK, Shamshi, MA & Sharma VP, 2000, 'Design Approach to use Platinum RTD Sensor in Snow Temperature Measurements', Journal of Instruments Society of India, vol. 30, no. 4, pp. 275-283, available at https://www.researchgate.net/publication/275276709

[7] Attri, RK, Sharma, BK & Shamshi, MA, 2000, 'Practical Design Considerations for Signal Conditioning Unit Interfaced with Multi-point Snow Temperature Recording System', IETE Technical Review, vol. 17, no.64, pp. 351-361, https://doi.org/10.1080/02564602.2000.11416928 or download from https://www.researchgate.net/publication/275276698.

Paper No.5

DESIGN OF AN INSTRUMENTATION SYSTEM TO RECORD DISTRIBUTION PROFILE OF SNOW LAYER TEMPERATURE FOR MODELING OF SNOW AVALANCHE FORECAST

RAMAN K. ATTRI

EX-SCIENTIST,
CENTRAL SCIENTIFIC INSTRUMENTS ORGANIZATION INDIA

Manuscript originally written Aug 1999

Abstract - The measurement of snow hydrological parameters is extremely important in developing a model for the prediction of snow avalanche and snow-melt water in the rivers. When direct measurement of these parameters is practically difficult, its dependence on snow temperature is used to develop snow cover models. A robust model for avalanche forecasting requires a sophisticated instrumentation system which can measure the required temperature parameters at right data points within the snowpack. A Snow Temperature Profile Sensing System along with surface temperature sensor has been designed to measure Snow temperature gradient, temperature distributions, and an average temperature of snowpack, snow surface, ground, and air. This paper describes the theoretical background to identify right temperature parameters needed to be measured and present a unique design approach to develop a measurement system to measure snow temperature at various points to provide an integrated data for the forecasting model.

1. INTRODUCTION

The snow manifests climatic changes, the potential release of avalanches, river run-off water, glacier sliding and related phenomenon in the mountain areas and planes nearby. The snow hydrological parameters and its dependence on other environmental factors govern the risk of snow avalanche and amount of water melting down the rivers (Yamazaki et al., 1993). Among all the environmental factors, temperature parameters directly affect the risk of the snow avalanche. It has been found that stability, strength, and structure of different layers composing snow pack mainly depends upon temperature distributions within and outside snowpack (Colbeck, 1989). If these temperature distributions are known, a hydrologist can develop a forecasting model for snow avalanche and generate a man-kind safety alarming system. However, the design of such alarming and forecasting system depends mainly on how accurately the right temperature variables are measured by an instrumentation system. The accuracy of snow cover avalanche forecasting model depends upon the selection of right temperature variables and right data points.

Researchers have developed a mathematical energy model of the snow cover (Singh, 1994; Yamazaki et al., 1993; Anderson, 1976). The major issue is to identify the right set of temperature parameters required for measurement and translating those variables into a feasible instrumentation system. In this paper, we have described the approach to identify the various temperature parameters required to be measured to develop snow cover avalanche forecast model and approach to design an instrumentation system to measure those parameters in physical form. Major focus of the paper is on how an instrumentation system can be design based on energy balance model of snow cover developed by Anderson (1976), Bader & Wielenmann (1992) and Singh (1994).

2. THEORETICAL BACKGROUND

Snow being a thermodynamically unstable material undergoes morphological changes such as crystal growth, change in density and internal weight pressure, percolation of snow-melt water, etc within the snowpack because of heat exchange at snow surface, ground interface and at air interface (Satyawali, 1994).

Many researchers established these morphological changes in the snow inherently depend upon the temperature parameters like - Snow layers temperature gradient, Average Temperature of Snowpack, Snow surface temperature, Ground interface temperature, and true air interface temperature (Schwerdtfeger, 1962).

Literature shows two approaches to establishing snow properties with temperature parameters:

2.1   Approach 1: Dependence of Snow Properties on Temperatures Parameters

The first temperature dependence of snow crystal growth on temperature has been shown by Mellor (1977) according to which rate of the morphism of crystals J depends upon the temperature gradient and snow surface temperature given by the following equation.

$$J = - \{(N\ D\ P\ L) / (RT)\} \{\partial T / \partial z\} \qquad .... (1)$$

The formations of snow cover take place with the development of different forms of crystals whose shape, size, bonding, and packing control the mechanical properties, stability and strength of snow cover thus evolved (Colbeck, 1989). The rate of this crystal growth is given $d_o = J/\rho$ where ρ is the ice density (0.917 gm/cm3. Using an expression from equation (1), the resultant crystal is grown because of the vapor deposition is given by:

$$d = d_i + M/T \{\partial T / \partial z\} \qquad .... (2)$$

Where M is a constant. The expression (2) shows that the final snow growth is directly proportional to the temperature gradient & inversely proportional to snow surface temperature.

Further, density and hence, snowbound mass is also related to this crystal growth which in turn depends upon temperature gradient of different layers. Snowbound mass is key component causing snow avalanche when excessive snow mass breaks the ground-snow bond underneath (Bader & Wielenmann, 1992). According to Mellor (1977) change in snow density depends exponentially upon average snow layer temperature:

$$\Delta\rho = A\ \rho\ Exp\ (-B\ \rho)\ Exp\ (-0.08(273- T_s))\ P\ \Delta t \qquad .... (3)$$

Where A and B are constants, ρ is the initial snow density (g/cm3), T_s is the average snow layer temperature in Kelvin, P is the overburden pressure and Δt is the time interval in hours.

Also, snow-melt depends upon the temperature distribution inside and above the snow. The study and forecasting of snow-melt water are necessary to forecast wet snow avalanche. The snow-melt produces some free water, the movement of this water vertically downward increases the rate of metamorphism, reduces mechanical strength and deformation resistance of snow pack by weakening inter-granular bonds (Anderson, 1976). With the increase in free water content, the snow tends to lose cohesion, and inter-granular bonds start to disintegrate. This causes a drastic decrease in shear strength and hence wet snow avalanche. The volumes of melt-water production for a given input approximately directly proportional to the snow-covered area and the temperature. Hall

(1985) states that energy input is frequently represented by degree-day factor and melt-water volume is calculated as follows:

$$V_m = a\,T\,A \qquad\qquad .... (4)$$

Where V_m is the melt-water volume in meter cubes, a is the degree-day factor, T is the number of degree days, A is the snow-covered area in meter squares.

Above evidence lead to the conclusion that snow surface temperature along with snow layer temperature distribution is a very important parameter needed for snow avalanche and flood run-off water forecast model.

2.2 Approach 2: Dependence of Heat Exchanges on Snow Cover on Temperature Parameters

It has been found by many researchers like Mellor (1977), Colbeck (1989), Anderson (1976) and Jordan (1991) that the stability, strength and other properties of snow are mainly related with the net exchange in at snow-air interface, snow-ground-interface and snowpack itself. The energy at the snow-air interface is considered to deduce the surface melt, energy exchange at the snow-ground interface examines the possibility of destruction of the bond between the ground and the snowpack. Net gains in energy, which is the directional addition of various energies at both the interfaces, satisfy the cold content, conduct through the snowpack and affect melt (Singh, 1994). A measurement of this energy enables hydrologists to make a snowpack behavioral model and hence develop an avalanche forecast model. These energies can be computed if physical parameters of snow surface temperature, temperature gradient etc is known.

Singh (1994) presented a snow cover model and depicted various energies invading on snow cover from the air-snow interface and snow-ground interface in figure 1. The snow-air boundary is being invaded by radiation energy Q_r (radiation flux) from above. Most of the energy at the snow-air interface Q_s (sensible heat flux) is used in satisfying cold content and changing the temperature of the top layer, some of the energy Q_l (Latent heat flux) produces a melt at snow surface, the rest of the energy is used in conduction at surface and absorption in interior layers.

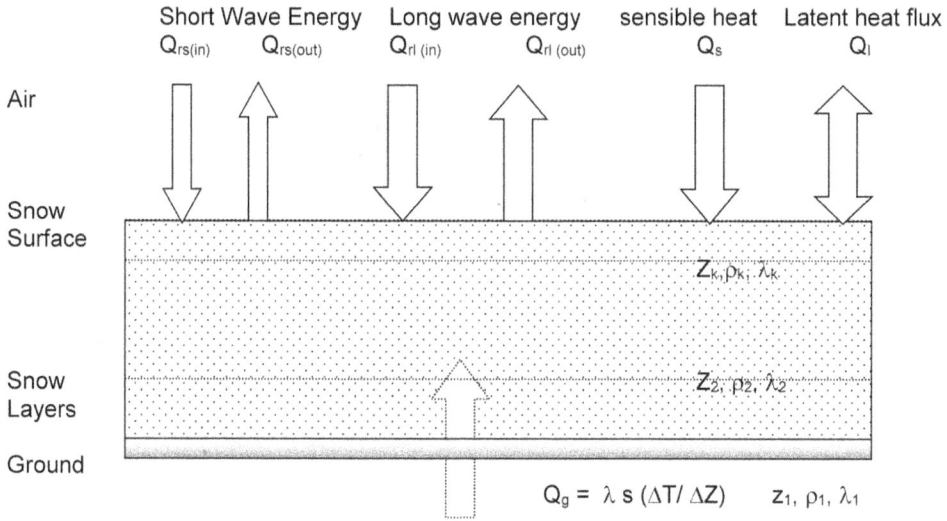

Fig. [1]: *Theoretical depiction of Snow Cover Model and surface energy exchanges at snow-air interface and snow-ground interface. An exchange of radiative energy as Short-wave and long-wave occurs at snow-air interface into the snow surface and out of snow surface. Sensible heat is absorbed which results in melting of the snow surface. Latent heat flux represents the energy used to melt F amount of snow at a depth of Z.*

Singh (1994) derived energy balance equation at snow surface governing the change in snow temperature as under:

Net Energy ΔQ = Radiation Energy Q_r + Sensible Heat Q_l + Latent Heat Q_s

$$.... (5)$$

Where ΔQ is the Net Energy Invading the snow surface, Q_r (radiation flux) is the radiation energy coming toward the surface, Q_s (Sensible heat flux) energy at the snow-air interface, Q_l (Latent heat flux) energy used in producing the melt.

Sensible heat Q_s is the sum of the two energy components namely the energy used in raising snow temp of snow depth Z and energy conducted into the snow surface. On the other hand, Latent heat is energy used in melting F amount of the snow of depth Z.

Above equation is expanded as under which is called snow-air interface energy balance equation:

$$\Delta Q = \varepsilon\, \sigma\, M\, [0.76T_a^4 - T_s^4] + C_s\, \rho_s \{dT_s/dt\}\, dz + \lambda_s\{\partial T_s/\partial z\}\text{ snow surface} + L_f\, F\, dz$$

$$.... (6)$$

Where T_a is ambient temperature in Kelvin, T_s is snow surface temperature in Kelvin, M is the constant of equation, C_s is the specific heat of snow, ρ_s is Snow density, dT_s is the change in snow temperature produced, z is the Snow depth, λ_s is the thermal conductivity of snow, $\partial T_s/\partial z$ is the temperature gradient in snowpack, L_f is the latent heat of fusion of snow, F is the amount of snow-melt per unit time and unit volume.

This is the basic equation which implies the direct dependence of energy exchange and hence meteorological interaction of snow-related to its temperature factors. Andersan (1976) and Ono et al. (1980) have further computed these individual terms in terms of snow surface and Air temperature etc. In the above equations, effects of all other parameters have not been included to keep the discussion to temperature distribution studies.

Schwerdtfeger (1962) proposed a Fourier one-dimensional heat conduction equation for snow cover which depicts heat conduction along the depth of snow Z as under:

Sensible heat flux at snow-ground interface Cs $\rho_s \{\partial T/\partial t\}$ = heat flux in ground λ_s $\{\partial^2 T/\partial z^2\}$

$$.... (7)$$

Above expressions establish these conclusions: -
(1) Radiative energy is a function of snow surface temperature and temperature gradient of the snowpack.
(2) Sensible heat depends on the temperature gradient of the snowpack and the time variation of the temperature.
(3) Latent heat depends on the snow cover thickness.
(4) Energy conducted within snow depends on the temperature gradient.

It has been established that if we could measure the following key parameters, we can directly compute above energies and hence the other interdependent properties of snow (Ganju, 1994).
- Ambient temperature T_a
- Snow surface temperature T_s
- Temperature gradient $\partial T_s/\partial z$
- Average temperature of snow pack, T_{avg}
- Temperature-time variation dT_s/dt
- Ground temperature T_g

A system has been designed to measure above key temperature parameters from which we can derive energy equations and hence other hydrological parameters. The following sections describe the design approach for developing a Snow Cover Temperature Profile Measurement System.

3. DESIGN OF SNOW TEMPERATURE MEASURING SYSTEM

3.1   Physical Design

To measure temperature distribution data, a Snow Cover Temperature Profile Measurement System has been designed. Proposed snowpack temperature profiler consists of a total of 29 temperature sensors distributed in 3D fashion along a 4-meter-long PVC rod. This system measures the temperature of different layers of snow at multiple points at different heights and different angles accurately. Also, the sensor for measuring the ground temperature at different depths inside the ground and true air temperature is a part of snow temperature profile sensor system (Shamshi et al., 1996).

The system requires great design efforts within tight environmental and operational specifications. The probe has to operate round the clock for several months under a temperature range from -50°C to +50°C with relative humidity up to 100% and the wind speed of the order of 200 Km/h. The system was required to be accuracy and a resolution of 0.1°C. In view of these severe environmental conditions, the packaging has been designed water and moisture proof & wires and cables are to be selected so that these can withstand these operational conditions. The components conform to military specification standard JM5555/JM38510-883 (Attri et al., 2000a).

3.1.1   Implementation of Snow Cover Temperature Profile Sensors: Measurement of T_{avg} and $\partial T_s/\partial z$

One-point measurement design cannot be implemented here because of different layers of snowpack having different properties, strength and energy contents and hence temperatures profile. We designed system architecture to support multi-point temperature measurements where the temperature of all the points is to be measured simultaneously to minimize the effects of time drifts. Care of putting enough points of measurement in snowpack has been taken (Attri et al., 2000a). One most important point is that points selected must be in a three-dimensional configuration.

A mechanical design of spiral profiler system with points of measurement lying on an imaginary cylindrical surface has been worked out. Refer to figure 2, in this configuration, multiple platinum RTD temperature sensors were spaced 30° apart in X-Z plane, 45° inclination downward of each of the mounted sensor in L-plane (cylindrical coordinates) and 20 cm apart in Y-axis from consecutive sensors. Sensor sealed in a metallic sheath is mounted on a triangular PVC assembly to give required slant of 45°. This slating angle made snow slide down & do not remain deposited over the probes during snow depletion. 30° horizontal spatial distances ensured that all directions around PVC mast are covered for temperature measurements. This arrangement gave overall 3-dimensional coverage of snow for temperature measurements.

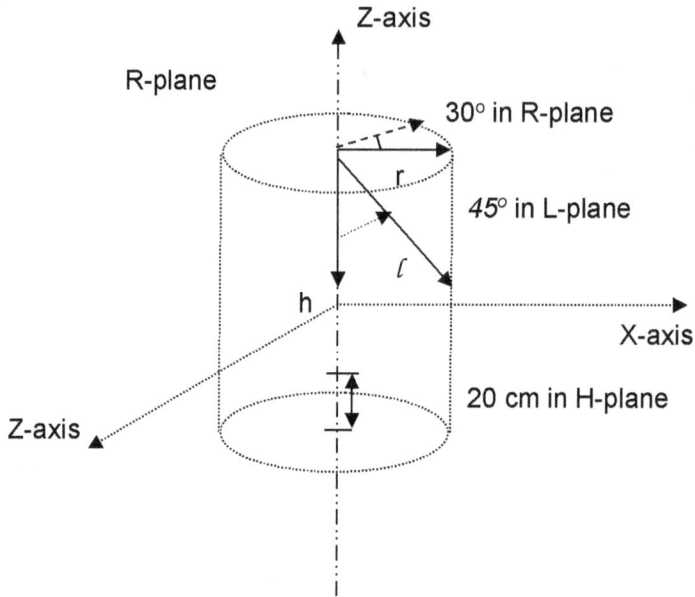

Fig. [2]: *Angular and spatial Orientation of Sensors in Cylindrical and Rectangular Planes. Each RTD sensor for snow cover temperature distribution measurement is mounted at a spacing of 20cm from each other in H-plane, mounted at 30 degrees angular spacing in R-plane from its adjacent RTD sensors giving progressive spiral coverage as seen from ground interface upwards. Each of the RTD sensors is mounted 45° slanted downwards to the Z-axis.*

Total 20 snow sensors have been used in the proposed snow temperature profile sensor system to cover approximately 4 meters of snowfall in a fashion shown in figure 3.

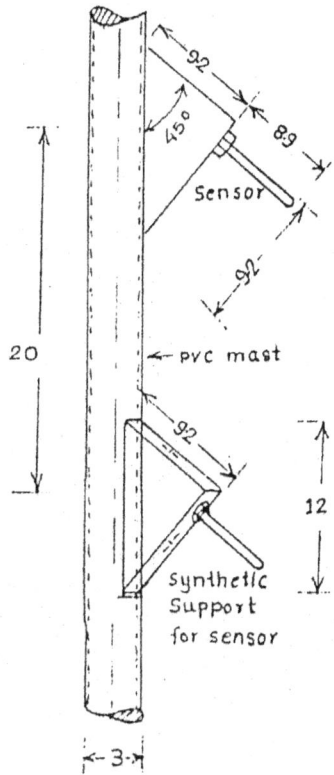

Fig [3]: *Enlarged View of a section of snow sensor PVC Mast. Each RTS sensor is mounted at an angle of 45 degrees with the help of a PVC triangular block flushed to PVC tube, spaced 20cm from each other vertically.*

The graphical view of these 20 sensors as seen from the top is shown in figure 4 with 30° separation and its physical implementation is as seen in figure 5. The leads from these sensors pass through the PVC tube and go to a junction box where the analog unit of the system is mounted. The junction box is mounted at such a height that it always remains out of snow cover. The analog output is passed through low-temperature cables to a data-logger (Attri et al., 2000a).

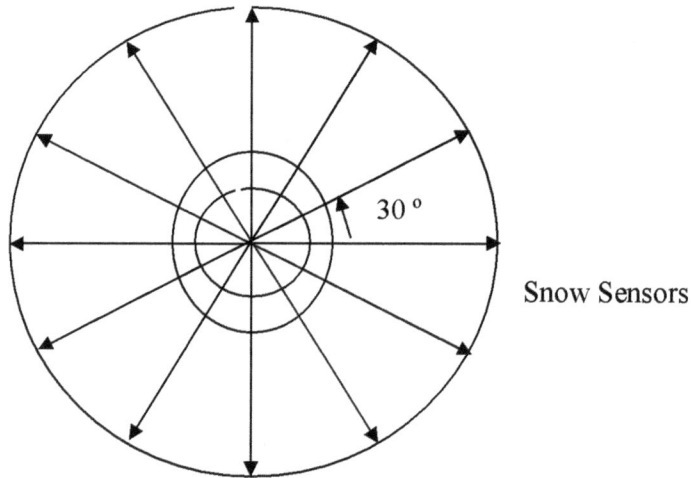

Fig [4]: *Graphical representation of angular mounting arrangements of RTD sensors as seen from the top of the PVC mast. Each RTS sensor block is spaced 30 degrees in R-plane from its adjacent RTD. This provides 360-degree coverage around the mast.*

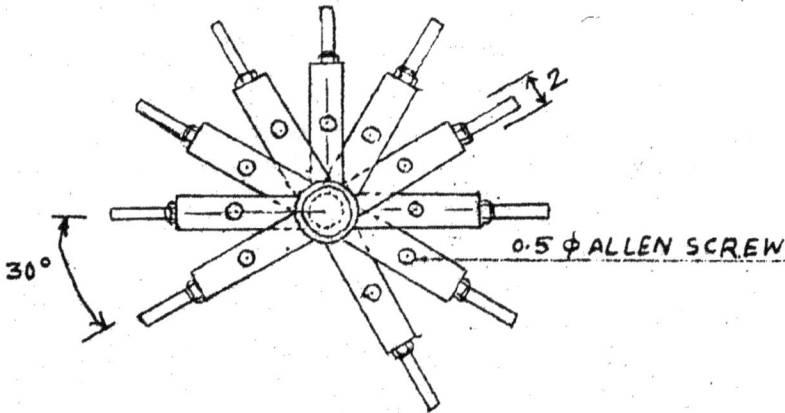

Fig [5]: *Physical mounting diagram depicting 30° Separation between 10 adjacent Snow Sensors in Spiral Assembly fitted on 2-meter PCV tube block as seen in top view. Each 2-meter segment of PVC masts carries 10 RTD sensor blocks. Two such segments are used to mount 20 RTS sensor blocks.*

3.1.2 Implementation of Ground Temperature Sensors: Measurement of T_g

Ground temperature is again important parameter to be measured. Here again layers of ground nearest to snow interface have different temperature as compared to the core ground layers. Here 3D data is not important. So, here two directional measurements are enough. Logarithmic variations of ground temperature have been considered, and hence the mounting of ground temperature sensors has been done accordingly. Total 7 number of ground sensors have been mounted. 1 m of a PVC tube buried inside the ground is mounted with seven sensing probes with tips coming outside at heights -5.0, -10.0, -20.0, -30.0, -50.0 and -100.0 cm relative to the ground. These 7 ground sensors provide a temperature profile of the ground and the temperature gradient at the ground interface. The figure 6 shows the ground sensors separations in –Z plane.

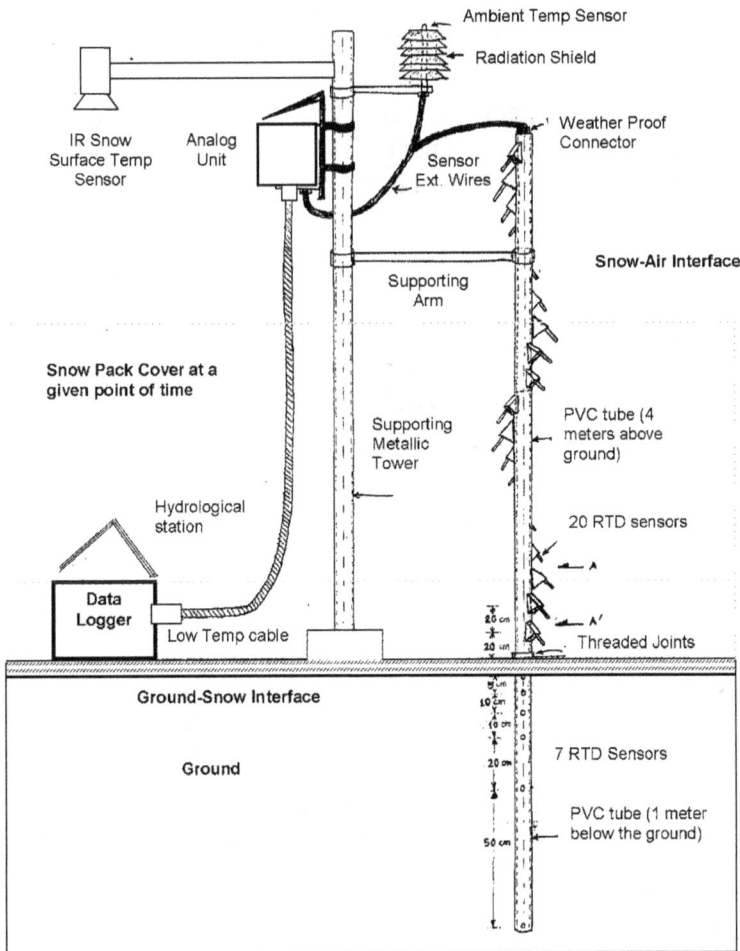

Fig [6]: *Complete mounting arrangement and physical arrangement of Snow Cover Temperature Profile Measurement System. 20 RTD blocks mounted on two segments of 2-meters each of PVC tube. 7 RTD sensors mounted on PVC tube of 1 meter in length as buds touching outside the periphery. Ambient temperature stays outside snow cover. One IR surface sensor is mounted facing downwards toward the snow surface.*

The installation arrangement of 29 sensors has been shown in figure 6. A total of 20 snow sensors have been mounted on the remaining 4m of PVC mast outside the ground at the spacing of 20 cm each slanting downward at 45 degrees angle and 30 degrees apart in the horizontal plane from adjacent sensors.

3.1.3 Implementation of Air Temperature Sensor: Measurement of T_{air}

One sensor is used to monitor the ambient temperature above the snow cover. This sensor is covered with a self-aspirated radiation shield as shown in figure 7 so that the sensor gives the true air temperature and is not affected by solar radiation. The air flows through the ventilation of this shield and raises the temperature of the sensing element. The heating caused by solar radiation and cooling caused by snowfall and rain does not affect the reading of this temperature sensor. Only the true air temperature is read. This sensor is to be mounted on the top of the PVC tube or the supporting tower. The ambient temperature sensor has the flexibility of increasing or decreasing the height of the sensor as the snow cover evolves or depletes off.

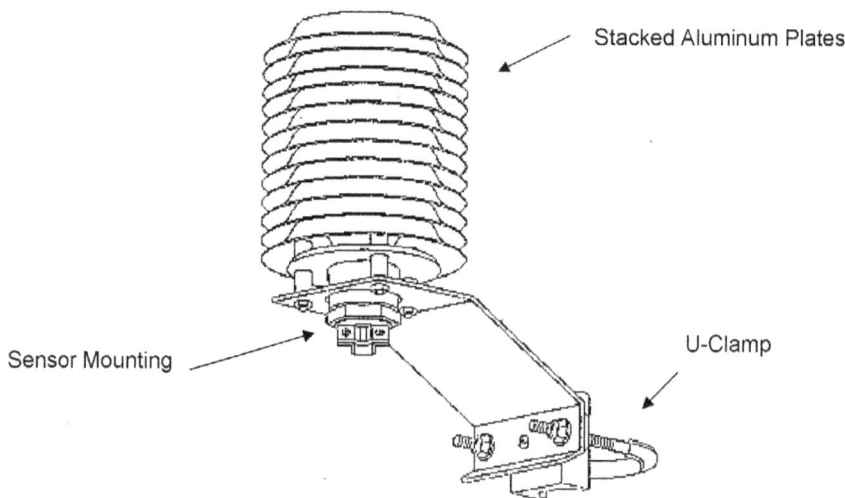

Fig [7]: Mechanical design of radiation shield for Snow air temperature sensor. The radiation shield is used to protect the RTS sensor from direct solar radiations and to ensure it measures true ambient air temperature.

3.1.4 Implementation of Snow Surface Temperature Sensor: Measurement of T_s

Snow surface temperature is a very difficult parameter to measure using any contact measurement method. The thermal radiation emitted by Snow has been exploited in the design of snow surface temperature sensor. A separate snow surface temperature sensor based on infrared technique has been designed as a separate independent unit and has been integrated with the above system. The infrared radiation emitted by the optically radiative snow is collected by the properly focused optical assembly. The temperature range of this sensor is from 0°C to −50°C with the resolution of 0.1°C in the measurement.

Tight moisture proof system has been designed in a single unit containing sophisticated optical assembly and the associated electronic circuit. Since the generated output signal is of very low level, signal conditioning unit is incorporated in the single unit inside the compact module along with the optical assembly. The unit is installed at some suitable height from the surface with the optical assembly facing vertically downward toward the snow. The system is powered by the data collection platform, which takes the readings of the snow surface temperature after some fixed interval. The readings are stored in memory for further processing.

3.2   Electronics Design

The complete signal conditioning circuits of all the sensors viz. snow sensors, ground sensors, and air sensor are contained in one single unit. This is a weatherproof sealed unit to withstand a low temperature. The complete block diagram of the system is as shown in figure 8.

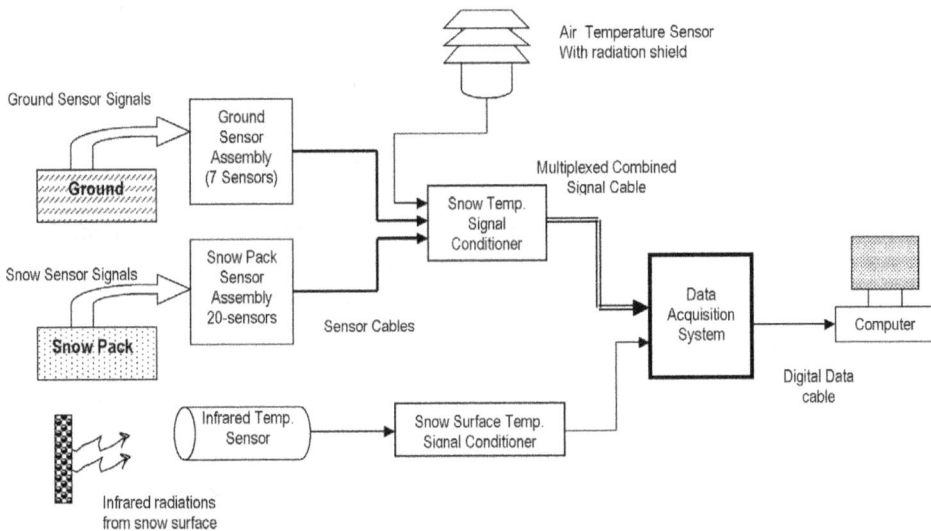

Fig [8]: A system block diagram of snow temperature profile measurement system. The 4-sensor assembly namely 20 snow cover temperature RTD sensors, 7 ground temperature RTD sensors, 1 ambient sensor, and 1 snow surface IR sensor. 28 RTD sensors have a common signal conditioning unit whereas IR sensor has its separate signal conditioning units. The output of signal conditioners is interfaced with a data-logger.

The principle of resistance variations with temperature has been used to measure Snow temperature. High accuracy, stable, precision metal film RTD sensor along with the associated circuitry has been used as the temperature sensing element. These sensing elements are enclosed in a metallic sheath filled with thermally conductive Aluminium Oxide. The compensation of self-heating, non-linearity, and extension lead error in the sensor has been provided in the electronic design of the system (Attri et al., 200b). The accuracy of the order of one-tenth of a degree Centigrade is required with a fair amount of linearity and stability.

The temperature sensed at 29 different points in a snowpack plus air interface temperature from the ambient sensor is stored in an electronic data acquisition system. It is a processor-based unit with all facilities of the modern high-performance data acquisition system. It controls the sampling interval and multiplexing of 28 channels. The multiplexed signal is given to A/D converter, which converts each analog value to the digital format. This digital data corresponds to temperature read by the corresponding sensor. This value is suitably computed by the processor and converted into an engineering unit. The value is stored in memory locations along with sensor identification number and time of recording.

The reading of temperature sensor is taken after selectable timing interval. After every hour, all the samples are computed to find out minima, maxima & average temperature of each sensor. The relative value in time gives the required parameter of the rate of change of temperature with time $\partial T/\partial t$. The computed data is important and is stored in a separate plug-in/plug-out memory module. Provision of data retrieval from a data-logger through the computer interface is also given. Data is downloaded by computer and used for further processing and modeling of the forecast. Data can be retrieved when required from these modules by reading them in memory reader in a laboratory.

Application software has been developed which plots temperature vs height of the sensor at a different time. This gives air temperature, snow temperature profile and ground temperature variations with time. This supports energy exchange computation & hydrological inferences.

The system is rigorously tested in the temperature range from −50ºC to +50ºC in the cold chamber.

4. APPLICATION OF MEASURED DATA USING SYSTEM

The Snow Cover Temperature Profile Measurement System has been installed in deep Himalayan regions (Shamshi et al., 1996). The Snow Temperature Profile Sensing System including ground temperature sensors, ambient temperature sensor, and the snow surface temperature sensor is used to get the hourly value of the average temperature of snow, average temperature of air, temperature gradient concerning the depth of snow and temperature variation with respect to time and average ground temperature.

Ganju (1994) presented following method to use physically monitored temperature data from the snow temperature profile sensor system to compute the energy exchange between air and snowpack and different hydrological parameters which depend mainly upon the temperature.

- Energy used in the cold content of the top snowpack is computed.
- The energy that can be conducted through the snowpack is computed using one-dimensional Fourier heat conduction equation with sources and sinks.
- The balance of energy is computed as subtraction of energy for cold content and energy conducted from net energy. A positive balance of energy used to calculate the snow melts.
- The next reading of snow temperature is taken after an hour. If snow temperature during the next hour is observed negative, the net gain in energy is once again used to satisfy cold content, raise the layer temperature and compute melt, if any.
- The Resultant snow layer thickness & density is computed based on the net melt.
- Compute wetting front depth for dry snowpack. When the snowpack becomes isothermal, melt-water is percolated. Once the melt-water reaches the bottom of the snowpack, the melt is taken out as subsurface run-off.
- Crust thickness is computed during the period of negative energy balance as there is no melt.

Using Armstrong approach, the type of metamorphism and the approximate size of grains are calculated. Using the model of the temperature profile, metamorphism and quantities computed above, a computer-aided simulation model of snow cover formation, thickness, density, melt, and strength is evolved which is incorporated in a snow avalanche forecast modeling, climatic forecast modeling and river run-off water determination modeling.

5. CONCLUSION

Data monitored by Snow Temperature Profile Sensing System has been successfully working round the clock in the Himalayan region. The basic snow avalanche forecast model was developed based on computer modeling of measured temperature data and using the equations of energy balance. However, mathematical models and equations used for extended computation and simulations are not solely dependent upon temperature parameters. The complete avalanche forecasting system requires a range of other sensors including snow depth sensor and humidity sensors. A comprehensive forecasting system is developed by monitoring snow depth and density, water contents, and total snow-covered area, morphological structure, and crystal growth rate, the temperature of warm rain and melt caused by it, reflectivity and absorption of solar radiation, porosity and strength, wind speed etc.

ACKNOWLEDGMENTS

Dr. M.A. Shamshi and Dr B.K. Sharma, Heads of Department, Geo-Scientific Instruments Division, Central Scientific Instruments Organization.
Mr V.P. Singh, Mrs Bupinder Kaur, Mr Rajender Shounda Technical officers, Geo-Scientific Instruments Division, Central Scientific Instruments Organization.

REFERENCES

[1]   Anderson EA (1976), A point energy and mass balance model of a snowcover, *NASA technical report*.

[2]   Attri RK, Sharma BK, Shamshi MA (2000), Practical Design Considerations for Signal Conditioning Unit Interfaced with multi-point Snow Temperature Recording System. *IETE Technical Review*, 17 (9): 351-61.

[3]   Attri, RK, Sharma BK, Shamshi MA, Sharma VP (2000) Design Approach to use Platinum RTD sensor in Snow Temperature Measurements, *Journal of the Instrument Society of India*, 30 (4): 275-283.

[4]   Bader HP and Wielenmann P (1992) Modelling temperature distribution, energy and mass flow in snowpack, *Cold Region Science and technology*, 20: 157-181.

[5]   Colbeck SC (1989), Snow crystal Growth with varying surface temperature and radiation penetrations, *Journal of Glaciology*, 35 (119): 23-29.

[6]   Ganju A (1994) Snow cover model, *Proceedings of SNOWSYPM-94*, 221-226.

[7]   Hall DK (1985) *Remote sensing of Ice and Snow*, London: Chapman and Hall Publications.

[8]   Jordan R (1991) A one dimensional model for snow cover, *CRREL Special report*.

[9]   Mellor M (1977) Engineering   properties of   snow, *Journal   of Glaciology*, 19: 15-66.

[10] Ono, WNM and Kawamura T (1976) Freezing phenomenon at sea water surface opening in polar winter, *Low temperature Science* 39: 159-166.

[11] Satyawali PK (1994) Grain Growth under temperature gradient, *Proceedings of SNOWSYPM-94*, 5-8.

[12] Schwerdtfeger P (1962) Theoretical derivation of Thermal conductivity and diffusivity of snow, *Int. Association of Scientific Hydrology*, General Assembly of Berkley, Belgium.

[13] Shamshi MA, Attri RK, Sharma, VP (1996) Snow Pack Temperature sensor, *Proceedings of National Conference on Sensors and Transducers*, 180-189.

[14] Singh AK (1994) Mathematical model for study of temperature profile within Snow cover. *Proceedings of SNOWSYPM-94*, 49-52

[15] Yamazaki, KTJ, Sakuraoka T and Nakamura T (1993) A one dimensional model of evaluation of snow cover characteristics, *Annual of Glaciology* 18: 22-26.

[16] Graya, MNT and Morland, LW (1994) A dry snow pack model, Journal of Cold Regions Science and Technology, 22 (2): 135-148.

[17] Anderson, EA (1976). A point energy and mass balance model of a snow cover. NOAA Technical Report NWS 19, Office of Hydrology, National Weather Service, Silver Springs, MD.

[18] Jordan, R (1996) A one-dimensional temperature model for a snow cover. Special report 657. US Army Cold Regions Research and Engineering Laboratory, Hanover, NH.

[19] Luce, CH and Tarboton, DG (2009) Evaluation of alternative formulae for calculation of surface temperature in snowmelt models using frequency analysis of temperature observations, Hydrol. Earth Syst. Sci. Discuss., 6, 3863–3890 http://www.hydrol-earth-syst-sci-discuss.net/6/3863/2009/hessd-6-3863-2009.pdf.

[20] Arons, EM and Colbeck, SC (1995).: Geometry of heat and mass transfer in dry snow: a review of theory and experiment, Rev. Geophys., 33, 463–493.

[21] Bartelt, P and Lehning, M (2002): A physical SNOWPACK model for the Swiss avalanche warning, 15 Part I: numerical model, Cold Regions Science and Technology, 35, 123–145.

[22] Jordan, R. (1991): A one-dimensional temperature model for a snow cover, Technical documentation for SNTHERM.89, US Army CRREL, Hanover, N.H. Special Technical Report 91-16, pp. 49.

[23] Koivasulo, H and Heikinheimo, M (1999).: Surface energy exchange over a boreal snowpack: Comparison of two snow energy balance models, Hydrol. Process., 13, 2395–2408.

[24] Tarboton, DG (1994): Measurement and Modeling of Snow Energy Balance and Sublimation, International Snow Science Workshop Proceedings, Snowbird, Utah, 260–279,

[25] Tarboton, DG, Chowdhury, TG, and Jackson, TH (1995): A Spatially Distributed Energy Balance Snowmelt Model, Biogeochemistry of Seasonally Snow-Covered Catchments, Boulder, Colorado, 141–155.

[26] Brun, E, Martin, VS, Gendre, C and Coleou, C (1989) An energy and mass model of snow cover suitable for operational avalanche forecasting. J. Glaciol., 35, 333–342.

[27] Bristow, KL and Campbell, GS (1984) On the Relationship Between Incoming Solar Radiation and the Daily Maximum and Minimum Temperature. Agricultural and Forest Meteorology, 31: 159-166.

[28] Colbeck, SC and Anderson, EA (1982) The Permeability of a Melting Snow Cover. Water Resources Research, 18(4): 904-908.

[29] Gray, DM and Male, DH (Ed.) 1981. Handbook of Snow, Principles, processes, management & use. Pergamon Press.

[30] Koivasulo, H and Heikenkeimo, M (1999) Surface energy exchange over a boreal snowpack. Hydrological Processes, 13(14/15): 2395-2408.

[31] Luce, CH and Tarboton, DG (2001) A modified force-restore approach to modeling snow-surface heat fluxes. Proceedings of The 69th Annual Meeting of the Western Snow Conference, Sun Valley, Idaho. http://www.westernsnowconference.org/2001/2001papers.htm.

[32] Tarboton, DG, Chowdhury, TG, and Jackson, TH (1995) A Spatially Distributed Energy Balance Snowmelt Model. In K. A. Tonnessen, M. W. Williams and M. Tranter (Ed.), Proceedings of a Boulder Symposium, July 3-14, IAHS Publ. (228): 141-155.

[33] Yen, YC (1967). The rate of temperature propagation in moist porous mediums with particular reference to snow. Journal of Geophysical Research, 72 (4): 1283-1288.

[34] Attri, RK (2018/1996) 'Snow Pack Temperature Profile Sensor,' R.Attri Instrumentation Design Series (Snow Hydrology), Paper No. 4, *Research and Design of Snow Hydrology Sensors and Instrumentation,* 2nd edn., pp. 59-64, Speed To Proficiency Research: S2Pro©, Singapore.

[35] Attri, RK (2018/2000) 'Design Approach to Use Platinum RTD Sensor in Snow Temperature Measurements,' R.Attri Instrumentation Design Series (Snow Hydrology), Paper No. 6, Research and Design of Snow Hydrology Sensors and Instrumentation, 2nd edn., pp. 66-75, Speed To Proficiency Research: S2Pro©, Singapore.

[36] Attri, RK (2018/2000) 'Practical Design Considerations for Signal Conditioning Unit Interfaced with Multi-point Snow Temperature Recording System,' R.Attri Instrumentation Design Series (Snow Hydrology), Paper No. 7, Research and Design of Snow Hydrology Sensors and Instrumentation, 2nd edn., pp. 66-75, Speed To Proficiency Research: S2Pro©, Singapore.

[37] Attri, RK (2018/2000) 'Design of A True Snow Air Temperature Sensing Probe,' R.Attri Instrumentation Design Series (Snow Hydrology), Paper No. 8, Research and Design of Snow Hydrology Sensors and Instrumentation, 2nd edn., pp. 66-75, Speed To Proficiency Research: S2Pro©, Singapore.

[38] Shamshi, MA, Attri, RK & Sharma, VP (1996) "Snow pack temperature sensor," Proceedings of National Conference on Sensors and Transducers, Chandigarh, pp. 180-189, viewed 24 Jan 2018, <https://www.researchgate.net/publication/275276742>.

[39] Attri, RK, Sharma, BK, Shamshi, MA & Sharma VP (2000) 'Design Approach to use Platinum RTD Sensor in Snow Temperature Measurements', Journal of Instruments Society of India, 30 (4): 275-283, available at https://www.researchgate.net/publication/275276709.

[40] Attri, RK, Sharma, BK & Shamshi, MA (2000) 'Practical Design Considerations for Signal Conditioning Unit Interfaced with Multi-point Snow Temperature Recording System', IETE Technical Review, 17 (64): 351-361, https://doi.org/10.1080/02564602.2000.11416928 or download from https://www.researchgate.net/publication/275276698.

DESIGN APPROACH TO USE PLATINUM RTD SENSOR IN SNOW TEMPERATURE MEASUREMENTS

RAMAN K. ATTRI

EX-SCIENTIST,
CENTRAL SCIENTIFIC INSTRUMENTS ORGANIZATION INDIA

The previous version of this paper was originally published as:
Attri, RK, Sharma, BK, Shamshi, MA & Sharma VP, 2000, 'Design Approach to use Platinum RTD Sensor in Snow Temperature Measurements', *Journal of Instruments Society of India*, vol. 30, no. 4, pp. 275-283, available at https://www.researchgate.net/publication/275276709

Abstract - The snow temperature measurement is a very critical area of hydrological instrumentation. The measured data is used to assess the run-off water, snowpack strength, and snow avalanche. The hydrological and snow avalanche forecast models require that temperature be measured at different points in the snowpack, above the snowpack and below the snowpack. A state-of-the-art multi-point snow temperature measurement system has been designed and developed for this purpose. The type of temperature sensor used is a critical aspect of this physical parameter instrumentation. After much study of various aspects of different temperature sensing elements, the platinum RTD PT-100 has been selected as a sensing element. This paper describes the selection criteria and performance evaluation of Platinum RTD sensor used in snow temperature measurement applications. The paper also highlights design techniques incorporated for minimizing the measurement error encountered while using Platinum RTD sensor.

1. INTRODUCTION

In the heavily snowbound areas, snow avalanche prediction is one of the most common as well as important forecast to be done. The forecast models depend upon the strength and physical properties of snow layers, snow deposition, and snow-melt parameters. The snow temperature profile directly governs these parameters. The information on multi-point snow temperature is very important for mathematical modeling and forecasting of avalanche formation, its release and discharge of water into rivers and stream due to melting. The accuracy in temperature measurement is very important in understanding the physics of avalanche formation and in forecasting of other related manifestations [1]. The modeling for forecasting requires very accurate measurement under stringent environmental conditions in the real-time mode and the required temperature measurement resolution is as minute as 0.1°C in the range from +50 °C to -50 °C. This imposes stringent and very critical design requirements over the measurement system. The temperature-sensing probe is the chief component, which decides the overall performance, accuracy, and linearity of the instrumentation system. The sensor used for measuring snow temperature must remain unaffected by the erroneous effects caused by the harsh environment. Therefore, the proper selection of sensing elements to be incorporated in the snow temperature probe is the prime design consideration.

Among the generally used sensors (thermistors, thermocouple, and RTD), the thermistor is one of the most commonly used temperature-sensing elements. Despite fairly high sensitivity and resolution, its resistance vs temperature characteristics are logarithmic and are not uniform over the entire range of required measurement [2]. On the other hand, the thermocouple is suitable for very high temperature measurements. In the negative temperature range, it does not work satisfactorily because metals need to be protected in moisture environment. Further, precautionary measures are to be incorporated for cold junction compensation, reference junction temperature control, and lead compensation. These measures make the measuring unit quite bulky and complex.

The RTD sensor has been found to have a fair degree of linearity as well as the stability of temperature measurement as compared to thermistors and thermocouples in the required temperature range. Thermistor and thermocouples have been found unsuitable for the present application as is clear from figure 1(a) and figure 1(b), where a comparison of linearity characteristics of the thermistor, thermocouple, and RTD has been given.

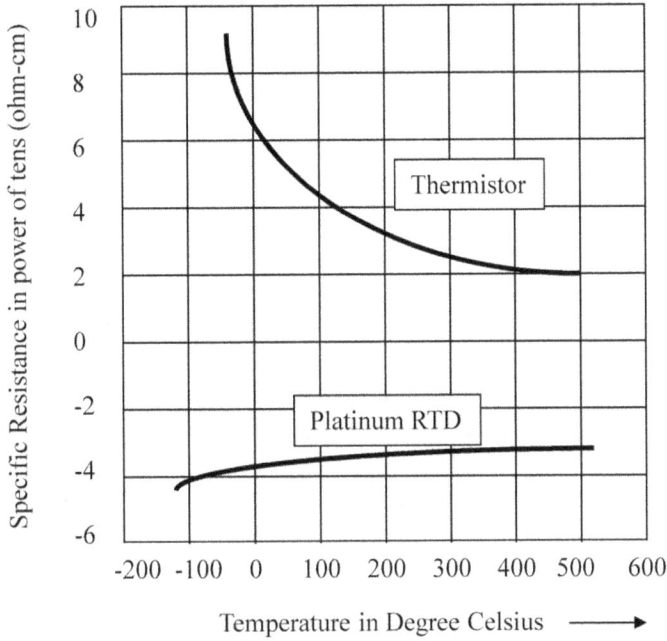

Fig [1] (a): *Comparison of temperature response curves of platinum RTD and thermistor*

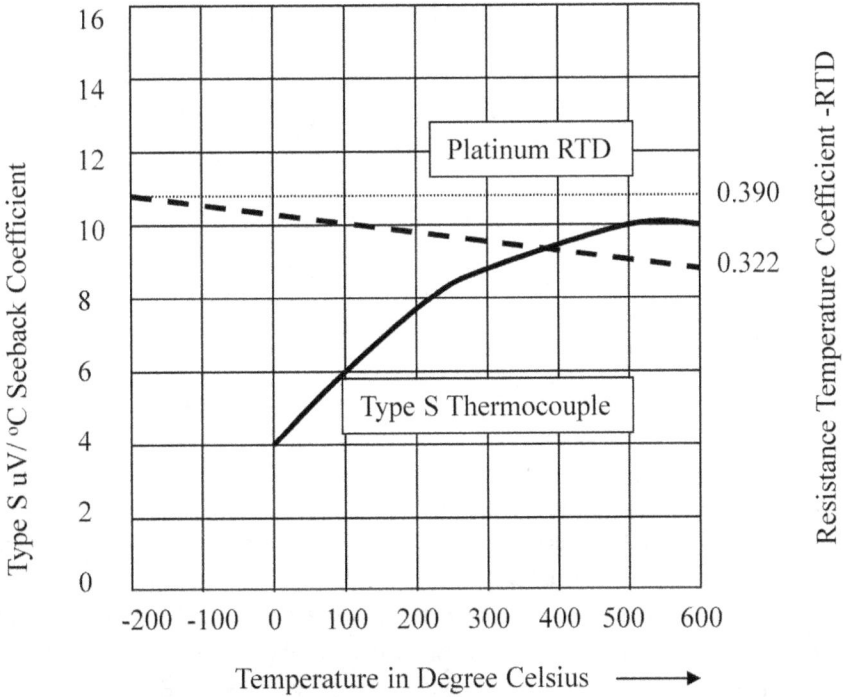

Fig [1] (b): *Comparison of equivalent non-linearity of thermocouple and RTD*

Thorough investigations on the basis of sensitivity, accuracy, linearity, error compensations and environmental conditions led to RTD as the appropriate sensing element in snow temperature profile measurement.

The paper describes the considerations involved in the selection of Platinum RTD sensor in the present application and its performance in snow temperature measurements. The design consideration of associated the electronic signal conditioner which removes the inherent errors in Platinum RTD and conditions the signal given by the sensor is also highlighted.

1.1   Using Platinum as the Sensing Element in RTD

In case of snow temperature profiler, the sensor has to be in contact with snow for a long time, and therefore the design of the sensor for environmental ruggedness is of prime significance. The specified working temperature range is from +50 ° C to -50 ° C. Linearity of the sensing element, as well as its fast response to minute changes in temperature is an important design consideration.

The performance of RTD is dependent upon the material used in its fabrication [3]. Commonly used materials in RTD are- platinum, copper, nickel, and tungsten. The design parameters [2] of these materials are shown in Table [1]. Since Copper does not work in the temperature range below 0°C , it is not suited to present applications. The platinum, nickel, and tungsten work fairly in the required range. But the range only is not the sufficient criteria, the resistance-temperature linearity plot is also to be taken into account.

Material	Temp. coeff.	Min. Range	Max. Range	Melting point
Platinum	0.39	-260°C	1100°C	1773°C
Copper	0.39	0°C	180°C	1083°C
Nickel	0.62	-220°C	300°C	1455°C
Tungsten	0.42	-200°C	1000°C	3370°C

Table 1: Parameters of Different temperature sensing materials for RTD [2]

In figure 2, the resistance-temperature characteristics of the different materials used in RTD fabrication have been compared. From the comparison, it is evident that only platinum has highly linear characteristics.

So, platinum has been found to be relatively linear within the specified range. The platinum further has additional merits, which make it suitable for the present application. These merits are- i) high precision and accuracy, ii) Ease of calibration, iii) High Repeatability, iv) fast response, v) Interchangeability with other resistance thermometers without any compensation, vi) Good performance in the desired temperature range, and vii) Limited susceptibility to contamination etc.

1.2   *Measurement Errors in Platinum RTD Sensor*

In spite of the merits of platinum RTD sensor, it has some inherent shortcomings which are to be overcome by adopting a better design strategy. These errors are mainly related to non-linearity and self-heating problems and affect the measurement to a greater extent because of highly tight specifications. The errors which are encountered while using platinum RTD sensors are- i) Self-heating error, ii) Extension lead error, iii) Thermal EMF error, iv) Non-linearity error, and v) Reference voltage variations error with temperature. All these errors have been eliminated by adopting suitable design strategy.

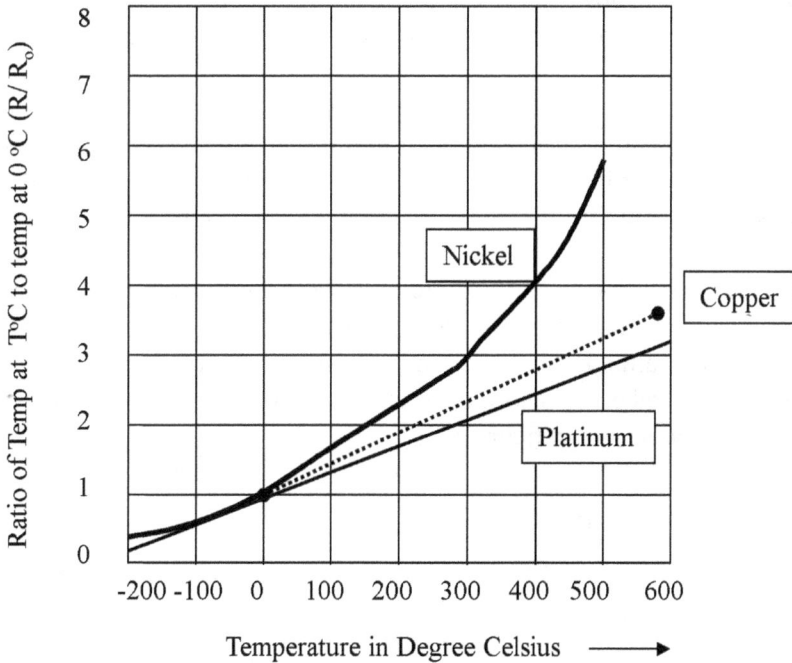

Fig [2]: *Comparison of resistance-temperature characteristics of different RTD materials*

2. *DESIGN APPROACH TO MINIMIZE THE ERRORS*

After giving due considerations to all the design factors, snow temperature profile measurement system has been designed and developed [5]. By evolving a new design approach, the above-mentioned errors have been minimized to meet the desired accuracy. The design methodology described herein gives the edge to RTD performance by removing all the observed errors.

2.1   Circuit Design for Reducing Self-heating Error

Unlike thermocouple, RTD is not self-powered. A current must be passed through the device to provide a voltage that can be measured. The current causes Joule(I^2R) heating within the RTD, changing its temperature. This self-heating appears as a measurement error. A typical value of self-heating error is 0.5°C per milliwatt in free air. An RTD immersed in a thermally conductive medium will distribute its Joule heat to the medium and the error due to self-heating will be smaller. The effects of the characteristics of the medium over the self-heating error are obvious from the fact that RTD, which causes an increase of 1°C per milliwatt in still air, will cause only 0.1 °C increase in temperature per milliwatt in the air which is flowing at the rate of one meter per second. In the present application, RTD being in contact with snow, the self- heating of RTD sensor has been found to cause snow-melt near it and thereby, changing the morphological state of snow. Hence the temperature read by the sensor will not be the true temperature of snow but of melted snow. It causes discrepancies in snow temperature profile curve.

In the present design, at first instance, the effort has been made for reducing the self-heating by emerging Platinum element in Aluminum Oxide paste which is a good thermal conductor and sealing it in a metallic sheath. The heat produced due to self-heating is instantly absorbed by aluminum oxide preventing the rise in temperature of the platinum element.

Further, the error due to self-heating is reduced by passing the minimum ohms measurement current that gives the desired resolution. If RTD is used in a bridge configuration, self-healing has been found to cause a great error. In the present case, a design has been worked out to incorporate a RTD sensor in the feedback path of negative feedback amplifier as shown in figure 3.

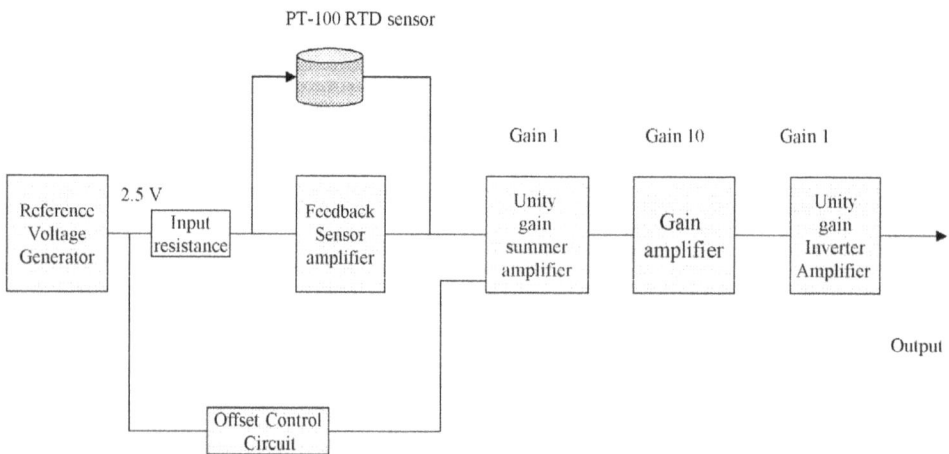

Fig [3]: Use of RTD in the feedback path of the amplifier to reduce self-heating

By designing in this manner, the current passing through RTD has been limited to the order of microamperes, thereby reducing the heating almost to a negligible value. The reference voltage source provides the stable voltage to the RTD connected in negative feedback amplifier path. The offset control circuit consisting of a fixed resistance and a variable resistance pot injects current into summing amplifier to give zero-volt output at $0°C$ at the final amplifier. This design scheme has provided the required resolution and has rendered the measuring probe very sensitive to detect small changes in the resistance and thereby sensing the variation in temperature.

2.2 Three Wire Configuration to Reduce RTD Extension Leads Error

The most unusual situation encountered in this case is the installation of the sensor at a large distance (4 meters in the present case) away from the signal conditioner module. This length of extension leads, connecting RTD junctions to the circuit, adds its own resistance to RTD path. Even if high-quality extension wires having the resistivity of 0.25 ohm per meter are used, 4 meter of wire in the circuit adds about 1 ohm of resistance in RTD path. This causes a severe error of the order of $3°C$ because 0.39-ohm resistance variation corresponds to $1°C$ error in temperature. In order to overcome this problem due to lead length, an offset control circuit has been incorporated in the circuit. Refer to figure 3 where offset control circuit design has been shown. It compensates the error caused by lead length by injecting an equivalent voltage in the summing amplifier.

Using the scheme as described here, the change in resistance of the extension wires due to temperature is compensated [6]. To nullify this error, three wires RTDs has been used in the feedback path as shown in figure 4.

The design configuration evolved in this way enables the system to respond only to resistance variation in RTD and eliminates lead error in conjunction with offset control. In figure 3, the resistance R_1 is input resistance while R_{11}, R_{12}, and R_{13} are the lead resistance of RTD. The lead resistance R_{11} in this arrangement gets added to large resistance R_1 and lead resistance R_{13} gets added to RTD resistance. The effect of adding R_{12} to the input impedance of the amplifier is negligible and hence can be ignored. In this case, the lead resistance gets distributed in three sections and gets added to another resistance in the path. For example, consider the circuit with R_{IN} equal to 10K ohm with RTD of 100 ohms in feedback. The gain of the circuit is 0.01. Now, if lead resistance R_{11}, R_{12} and R_{13} are 0.23 ohm each, the resultant value of R_{IN} and RTD will be 10000.23 ohms and 100.23 ohms. The resultant gain shall be 0.01002, and the effect of adding lead resistance is not coming into the picture. The third lead resistance 0.23 ohm get added into input impedance (of the order of 1M ohm) of the amplifier and its effect is also negligible. The net result of using this configuration is the elimination of the effect of lead resistance.

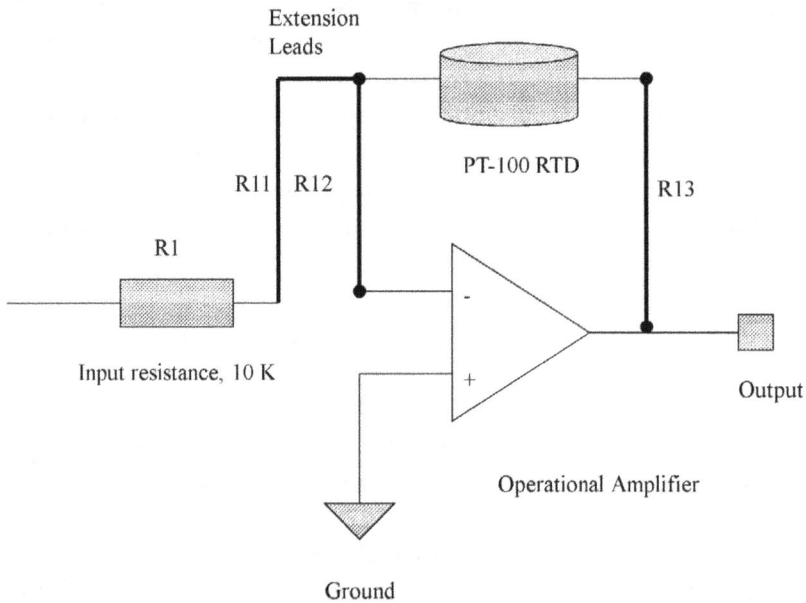

Fig [4]: *Use of three wire RTD to reduce extension lead error*

2.3   Offsetting Thermal EMF Error

The platinum-to-copper connection made during fabrication of RTD causes a thermal offset e.m.f. Voltage due to the phenomenon of Seeback effect (i.e. an e.m.f is produced at the junction of two different metals). The offset-compensated ohm technique has been used to eliminate this effect. When the total system is integrated, the net output offset voltage will comprise of i) offset due to EMF, ii) offset caused by extension leads and iii) inherent offset of the operational amplifier. All these offsets voltages are difficult to differentiate from each other. The circuit, which has been designed for eliminating error due to extension leads, eliminates this collective offset voltage as well.

2.4   Linearizing Resistance-Temperature Curve of RTD

From the resistance-temperature curve of platinum in the range of -50°C to +50°C, it is inferred that the dispersion of resistance is not equal for equal dispersion in temperature on both sides around 0°C.

Temp in °C	Resistance (ohm)	difference
-50	80.25	0.40
-49	81.65	0.40
-48	82.05	0.39

0	100.00	0.39
+48	118.62	0.39
+49	119.01	0.38
+50	119.40	0.39

Table 2: Excerpt from Resistance-Temperature Table (European Curve, Alpha=0.00385) of Platinum RTD [4]

The resistance of platinum at -50 °C is 80.25 ohm (refer to resistance-temperature table 2) which is a dispersion of 19.75 ohms from the zero-degree value of resistance (100 ohms). Whereas on the other side at +50 °C, the resistance value is 119.40 ohm, which is a dispersion of 19.40 ohm from zero temperature value. Therefore, a net difference of 0.35-ohm (19.75-19.40) in resistance dispersion at end-points (-50°C and +50°C) of required range on both sides 0°C. This is the deflection in platinum resistance curve obtained in the range of temperature from linear ideal curve.

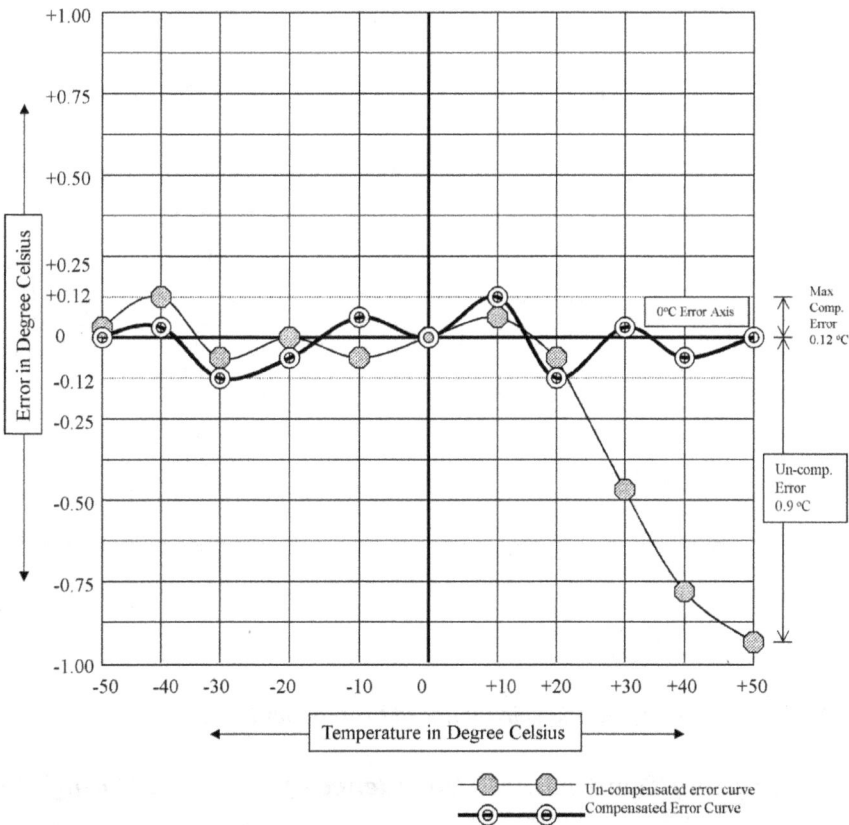

Fig [5]: Un-compensated and compensated response curve of Pt-100 RTD

The temperature coefficient of platinum is 0.39 ohm/°C and the above deflection of 0.35 ohm indicates a net error of 0.9°C in platinum temperature response on either side of the range of -50°C to +50°C. This defection is plotted in figure 5 depicting non-linearized response of platinum.

As is clear from the figure 5, this non-linearity error is severe. To tackle the non-linearity problem, a gain control circuit has been designed to change the gain in positive and negative range with proper switching at some suitable reference point. The gain control circuit automatically changes the gain in positive range after comparing with a suitable voltage reference. In the present case, the reference has been taken as 10°C point onward from where non-linearity has been observed to start. This design configuration as shown by block diagram in figure 6, further gives four-point calibration of the system providing increased accuracy and improved linearity. The four points for calibration are- 0°C point, - 50°C point and +50°C point and +10°C reference point. The output of the circuit is plotted against temperature after calibration against these four points as shown in figure 5. This gives the final performance curve of platinum, which is highly linearized, and accurate within the required specification. Maximum dispersion at any point does not exceed 0.12 °C.

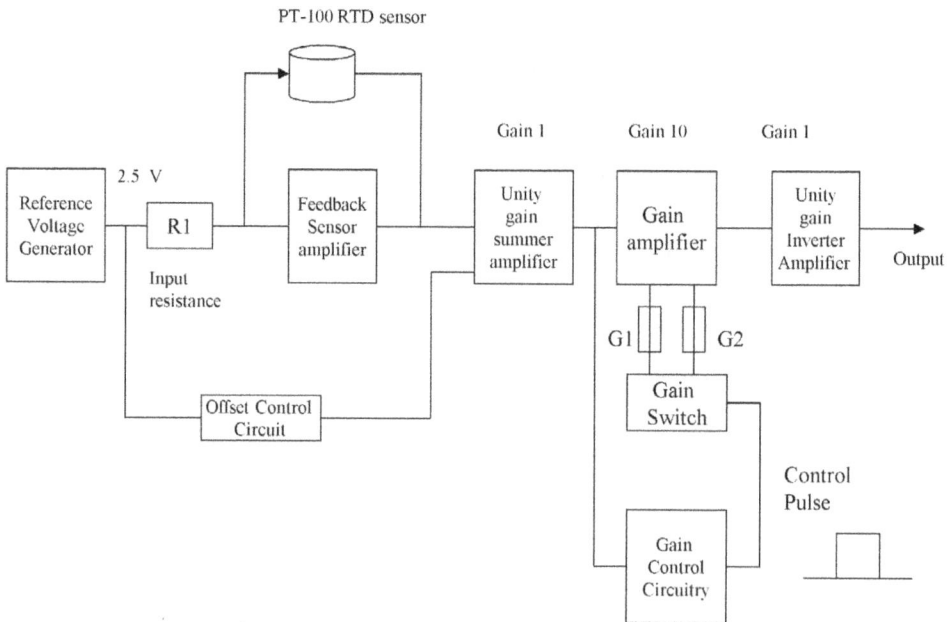

Fig [6]: Non-linearity compensation circuit added to amplifier block

2.5   *Eliminating Temperature Dependence of Reference Voltage Error*

An error has been observed because of the changes in the reference voltages due to temperature. This is reflected as if the temperature of the RTD sensor has changed. To overcome this problem, a highly stabilized reference voltage generator has been designed with a very small temperature coefficient. The average temperature coefficient of this reference voltage regulator is 0.002 %/°C in the range of -55 °C to +125 °C. And long-term stability is 0.1%/1000 hrs. Using military specification ICs and thin film 0.1% resistance and Stereo-flex/Tantalum capacitors etc, the results have been improved remarkably

3. OTHER DESIGN CONSIDERATIONS

Besides the techniques mentioned above, some practical precautions are to be taken into consideration. These are related to using of shields and twisted-pair wire, use of proper sheathing, use of guarded integrating DVM, and avoidance of stress and steep gradients etc. In addition, precautions related to size and construction have been observed. Due to its construction, the RTD is somewhat more fragile than the thermocouple and precautions must be taken to protect it [7]. For that, the sensing element is sealed in metallic sheath filled with a thermally conductive material like Aluminum Oxide. For reducing the self-heating, both the size and the current as well has to be considered. Small RTD will be having a fast response time and low thermal shunting but will give high self-heating error. On the other hand, large RTD will inhibit response time but will reduce the self-heating error. Minimum ohm measurement current and larger RTD (larger size gives a better reduction in self-heating) which can still give good response time, has been used. But care has been taken not to increase the size of the sheath too much just in order to reduce self-heating.

For better response, Metal Film RTD's can be used. These elements are constructed by depositing or screening platinum or metal-glass slurry film onto a small flat ceramic substrate, etched with laser –trimming system and sealed. These film RTDs has the advantage of increased resistance for a given size. Due to small size, it can respond quickly to step changes in temperature. The measurements are decidedly improved by these RTD's.

4. CONCLUSION

It is obvious from above discussions that accuracy, linearization, performance and stability of the temperature measuring system depends primarily upon the sensor performance and remedial measures taken against any error arising therein. Because of its use in a harsh environment, the performance of the sensor has been critically analyzed. Most of the errors have been reduced to a greater extent to obtain a linearized response free from errors due to self-heating and lead length. The designed Snow Temperature Profile Temperature Sensor system is operated in highly snowbound areas round the clock throughout the year and collects the accurate data of snow temperature profile within the snowpack, which is stored in the interfaced data acquisition system. The performance of the system designed on the above design methodology has been found very satisfactory in the desired negative and positive temperature range. The system has been designed in such a manner that the system can also be used for monitoring the minute variations in temperature measurements of oil, water hotness, and ambient conditions.

REFERENCES

[1]   Singh, AK, 1994, Mathematical model for study of temp profile within Snow Cover, Proceedings of SNOWSYPM-94, pp. 49-52.

[2]   Temperature Measurement Handbook, Omega Engg Inc, Stanford, Connecticut.

[3]   Rosemount, 1962, Platinum Resistance Temperature Sensors, Bulletin 9612, Rosemount Engg Co.

[4]   Platinum RTD Resistance table, Omega Temperature Catalogue, Vol 26, Omega Engg Inc.

[5]   Shamshi, MA, Attri, RK, Sharma,VP, 1996, Snow Pack Temperature sensor, Proceedings of National Conference on Sensors and Transducers, pp. 180-189.

[6]   Evans, JP and Burns, GW, 1962, A study of stability of High temperature platinum Resistance Thermometer (Temperature-its measurements and Control in Science and Industry), Reinhold, NY.

[7]   White, GK, 1979, Experimental Techniques in Low Temperature Physics, 3rd Edition, Clarendon Press, Oxford.

[8]   Attri, RK 2018/1996, 'Snow Pack Temperature Profile Sensor,' R.Attri Instrumentation Design Series (Snow Hydrology), Paper No. 4, *Research and Design of Snow Hydrology Sensors and Instrumentation,* 2nd edn., pp. 59-64, Speed To Proficiency Research: S2Pro©, Singapore.

[9]   Attri, RK 2018/1999, 'Design of an Instrumentation System to Record Distribution Profile of Snow Layer Temperature for Modelling of Snow Avalanche Forecast,' R.Attri Instrumentation Design Series (Snow Hydrology), Paper No. 5, Research and Design of Snow Hydrology Sensors and Instrumentation, 2nd edn., pp. 66-75, Speed To Proficiency Research: S2Pro©, Singapore.

[10] Attri, RK 2018/2000, 'Practical Design Considerations for Signal Conditioning Unit Interfaced with Multi-point Snow Temperature Recording System,' R.Attri Instrumentation Design Series (Snow Hydrology), Paper No. 7, Research and Design of Snow Hydrology Sensors and Instrumentation, 2nd edn., pp. 66-75, Speed To Proficiency Research: S2Pro©, Singapore.

[11] Attri, RK 2018/2000, 'Design of A True Snow Air Temperature Sensing Probe,' R.Attri Instrumentation Design Series (Snow Hydrology), Paper No. 8, Research and Design of Snow Hydrology Sensors and Instrumentation, 2nd edn., pp. 66-75, Speed To Proficiency Research: S2Pro©, Singapore.

[12] Shamshi, MA, Attri, RK & Sharma, VP 1996, "Snow pack temperature sensor," Proceedings of National Conference on Sensors and Transducers, Chandigarh, pp. 180-189, viewed 24 Jan 2018, <https://www.researchgate.net/publication/275276742>.

[13] Attri, RK, Sharma, BK, Shamshi, MA & Sharma VP, 2000, 'Design Approach to use Platinum RTD Sensor in Snow Temperature Measurements', Journal of Instruments Society of India, vol. 30, no. 4, pp. 275-283, available at https://www.researchgate.net/publication/275276709

[14] Attri, RK, Sharma, BK & Shamshi, MA, 2000, 'Practical Design Considerations for Signal Conditioning Unit Interfaced with Multi-point Snow Temperature Recording System', IETE Technical Review, vol. 17, no.64, pp. 351-361, https://doi.org/10.1080/02564602.2000.11416928 or download from https://www.researchgate.net/publication/275276698.

Paper No.7

PRACTICAL DESIGN CONSIDERATIONS FOR SIGNAL CONDITIONING UNIT INTERFACED WITH MULTI-POINT SNOW TEMPERATURE RECORDING SYSTEM

RAMAN K. ATTRI

EX-SCIENTIST,
CENTRAL SCIENTIFIC INSTRUMENTS ORGANIZATION INDIA

The previous version of this paper was originally published as:
Attri, RK, Sharma, BK & Shamshi, MA, 2000, 'Practical Design Considerations for Signal Conditioning Unit Interfaced with Multi-point Snow Temperature Recording System', *IETE Technical Review*, vol. 17, no.64, pp. 351-361,
https://doi.org/10.1080/02564602.2000.11416928 or download from
https://www.researchgate.net/publication/275276698

Abstract - Multi-point temperature measurement has always been a very important aspect of physical instrumentation mainly in environmental and industrial application. A multi-point temperature measurement system has been developed to measure the temperature at different points simultaneously. This multi-channel system is designed specifically for measuring snow temperature at 28 different points in snow layers and this multi-point data is used in modeling of snow cover. The interfacing of such multi-channel system to data acquisition system is one of the most critical design aspects, which affect the overall performance of the system. The temperature sensor system and data acquisition system may be working satisfactorily when operated independently but total system integrity is absolutely important when these two sub-systems are interfaced together. Further operating conditions in the snow cause many performance and reliability issues to be tackled through proper design. This paper discusses the design aspects involved in the interfacing of the multi-channel temperature measurement system with the data acquisition system specifically for Snow hydrological applications.

1. INTRODUCTION

The temperature is an extremely important parameter to be measured in environmental, Hydrological and other industrial applications. A temperature sensing system has been designed specifically for snow hydrological applications and related forecast [1]. The signal generated from this sensor system is properly processed and stored in a suitable format, which is further, downloaded by computer. This data is then used in hydrological studies and forecast models. Since the snow is deposited layer by layer, so a multi-point measuring system is specifically needed here to measure the temperature of each layer as per the requirement of the hydrological models [2,3,4,5]. A suitable design has been worked out to measure the temperature of snowpack at 28 different points in 3-dimensional coverage of temperature distributions [1]. Total 28 temperature sensors have been used in the measurement at these 28 points situated at various depths of snow.

This paper discusses the major design aspect of interfacing such a multi-point temperature measurement system to a data acquisition system. Further, the factors affecting the performance and reliability of the measured signal has been discussed.

2. SYSTEM OVERVIEW

The system consists of two self-contained units. One is signal conditioning unit (SCU) and other is data acquisition system (DAS). Signal Conditioner contains the sensor interface, analog, and measurement circuitry. The platinum RTD sensing element has been used for the measurement of temperature at different points [6,7,8]. Compensation of non-linearity and self-heating has been provided [9,10]. The RTD has been used in the feedback path of the op-amp to reduce the self-heating error. The single point temperature measurement system, as shown in figure 1, includes a reference voltage generator, sensing amplifier, gain amplifier, and final buffer amplifier. PT-100 platinum RTD with a sensitivity of 0.39 ohm per degree Centigrade temperature has been used in the feedback path and the gain of this amplifier changes in accordance with the variation in the resistance of the RTD, which in turn is proportional to the temperature. The voltage output at the output of the amplifier changes proportionally to the temperature [10].

The precise relationship connecting the resistance and temperature is given by Calendar Relation as follows:

$$R_t = R_o + R_o \, \alpha \, [t - \delta \, (t/100 - 1) \, t/100]$$

Where
$\alpha = 0.00392$ and $\delta = 1.49$

This relationship can be approximated for $-50°$ C to $+ 50°$ C temperature range as:

$$R_t = R_0 (1+ \alpha\ t)$$

The current proportional to offset is added into the summing amplifier to get zero output at 0°C. A voltage of 33-mV and 100-mV per $^{\circ}$C change in temperature is obtained at the summing amplifier and final amplifier output respectively. The performance of the design has been very reliable and up to the accuracy of 0.1 $^{\circ}$C. The voltage proportional to temperature is read by a data acquisition system (DAS) through interfacing cable. This reading is converted into a digital value by ADC and stored in memories along with auxiliary information. This data is further downloaded or transmitted to a PC or central station for analysis. DAQ system hardware has been designed around microprocessor with 16-bit ADC700 Analog to Digital Converter chip having a resolution of 156μV. The system is software controlled with in-built EPROM operating software controlling number of channels, ADC conversion, formatting of digital data and data stored in memory. The software further controls RS232 data communication interface for data downloading into PC.

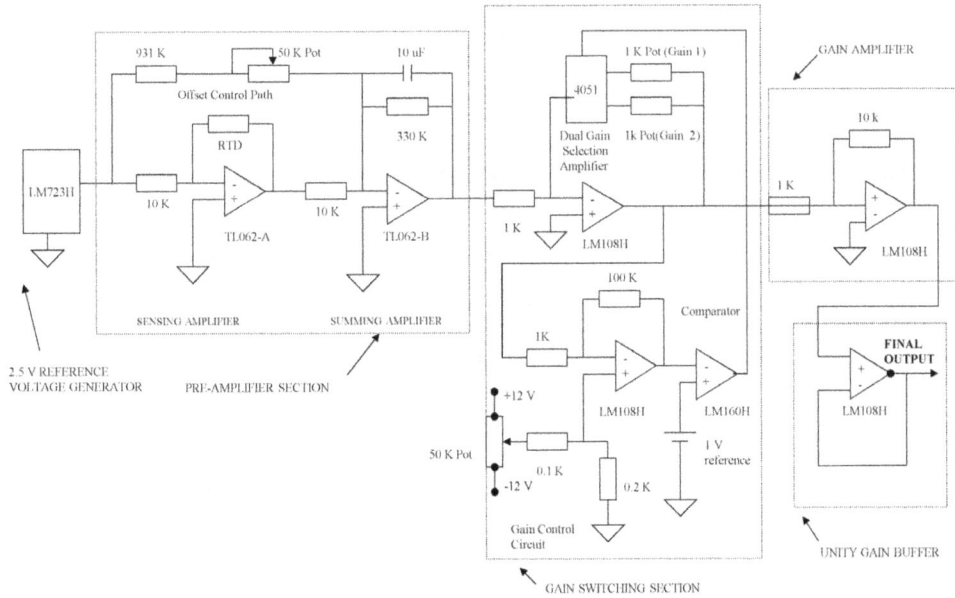

Fig [1]: Single-channel circuit diagram of snow temperature measurement system

3. DESIGN RESTRICTIONS

There are certain tough environmental as well as design restrictions on the system. The design strategy for above single-channel system has to be extended to implement the 28-channel system in such a way to make a portable weatherproof field operated unit with minimum current consumption. 28 sensing elements are to be used and their respective signal is to be processed in the same signal conditioning unit. These 28 points have been selected in a group of 8, 10 and 10[1]. The number of components used has to be minimum in view of little availability of military grade ICs and components and to make a portable compact system. This is to be done in order to reduce current consumption of the system since the system has to work in remote areas and has to operate on batteries. The interfacing cable from the signal conditioning unit to the data acquisition system is about 50 meters, so the voltage drop factor must be considered. The overall system should operate in the temperature range from -50 °C to +50 °C and overall assembly has to be totally weatherproof.

System design conforming to these tough specifications has been worked out and shall be discussed here. The output of the 28 channels has to be connected to Data Acquisition through a 50-meter cable. There are many practical problems in carrying independent outputs of each channel through a multi-core cable to such a long distance. To overcome this problem, a 28-channel multiplexer is inserted in the circuit at a suitable point. The multiplexer will combine the outputs from different channels onto a single output line [11]. The first major consideration in this design is to select the insertion position of the multiplexer so that the minimum number of components is used. The second major design consideration is the propagation delays and settling time involved. And third design consideration is noise and grounding aspects involved.

4. *DESIGN APPROACH FOR MULTIPLEXING OF CHANNELS*

As related to the position of the multiplexer in the circuit, there are two obvious positions. First one is the usual configuration after the final buffer amplifier. This configuration will have the advantage of completely independent channels with individual gain and non-linearity settings giving a performance improvement. But the total number of ICs used in the complete system shall be:

1 (common reference voltage source) + 6 (IC in one circuit) x 28 + 9 (multiplexer circuit) = 178 ICs.

An alternative suitable position is to introduce multiplexer between pre-amplifier and the gain control circuit and it has been found that the system performance is fairly acceptable. This configuration results into a system with:

1 (reference source) + 1(pre-amp) x 28 + 9 (MUX circuit) + 5 (gain scaling circuit) = 43 ICs.

In this configuration, the size of the unit is reduced to one-fourth resulting in a compact small sized 28-channel unit. Further the current is reduced to one-eighth as compared to the above configuration.

Form Factor Improvement   = 400 %
Current Reduction   = 800 %

This advantage definitely makes the system attractive for applications in the remote areas where the power to the system is supplied by the battery. Only the individual setting of each channel is the offset voltage setting at zero degree Centigrade.

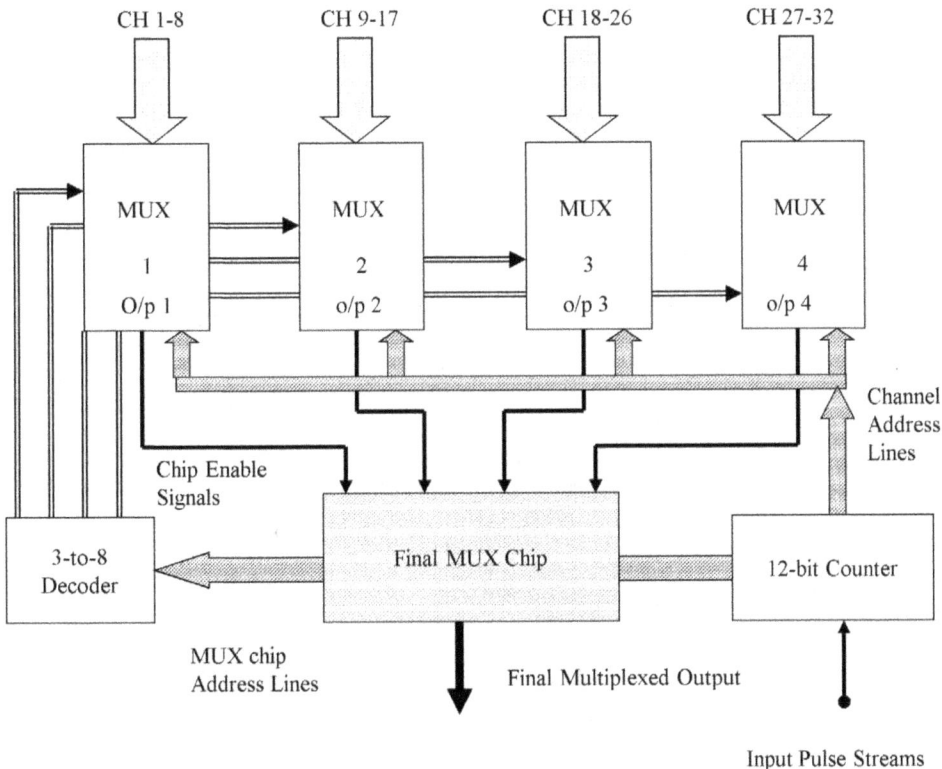

Fig [2]: *Multiplexer Circuit Block Diagram*

The first step is to design the multiplexer circuit for multiplexing 28 channels onto a single line. It has been worked out that the multiplexer circuit has to be kept on the signal conditioner side instead of conventional design of keeping the multiplexer on the side of data acquisition system. Only this design strategy shall send the output of the channels pre-multiplexed over one line in the signal conditioner under the control of DAS. The multiplexer chips like CD4051 with 8 input lines are readily available. In order to integrate 28 inputs, 4 multiplexer CD4051 are used in series. The 4 outputs of these multiplexers further integrated into another multiplexer chip CD4051 to get one multiplexed output line. A stream of 28 pulses (Channel Select Pulses) from DAS is used to select the individual channels one after another. A control circuit has been designed which provides the proper selection of channels to the multiplexers from these pulse streams. This control circuit consists of a 12-bit counter (CD4040) and a 3-to-8 decoder (74HC902). The output from the selected channels is read by following circuitry consisting of a gain switching section, gain amplifier and buffer amplifier after a suitable time depending upon propagation delay and settling times of operational amplifiers. The 28-channel system designed on these lines is shown in the figure 2.

The single-channel circuit described earlier is extended to design multi-channel system by introducing the multiplexer circuit after pre-amplifier stages as shown in figure 3.

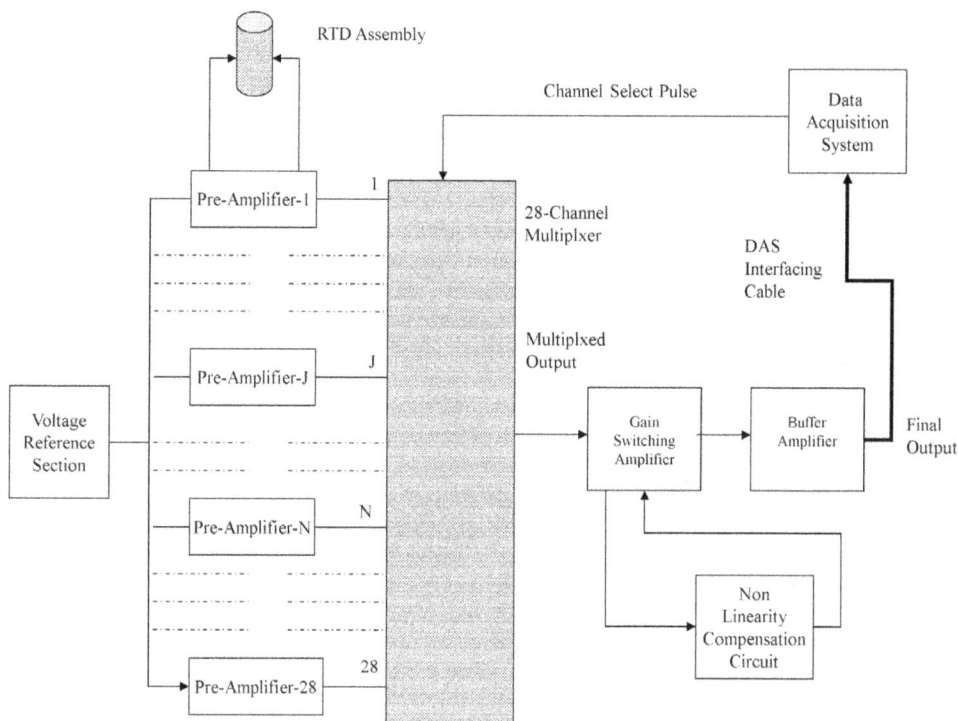

Fig [3]: *Multi-channel temperature measuring system block diagram after inserting multiplexer circuit between pre-amplifier stage and gain control amplifier stage*

5. MEASURING SYSTEM INTERFACING ASPECTS

The signal conditioner circuit is tested and calibrated independently using ideal signals for 32 Pulse streams. The offset control is used to calibrate 0°C point and gain control is used to calibrate the system at −50 °C and +50 °C point. Similarly, DAS is also tested independently with an ideal input voltage source. When these units are to be interfaced together to give the best possible performance, many practical errors come into picture which is removed through proper design discussed in the following sections.

A high input impedance buffer amplifiers configuration with unity gain has been designed as the final stage of the Temperature measuring unit as well as the input stage of DAS as shown in figure 4. This serves as the isolation shield between the signal conditioning circuit and DAS and further performs the function of impedance matching [12]. These two buffers in combination minimize the impedance reflections on both sides of the interface. Proper voltage follower driving amplifiers (LM102H in this case) with more than 10^{12} ohms input impedance and 0.8-ohm output impedance are selected which could respond well to rapid changes in input current as the ADC conversion proceeds. Loading of analog output by the ADC has to be specifically taken into account. Since the tri-state ADC (ADC700) has been used, it offers different impedance to the signal in the active mode (about 5 k ohm) and tri-state mode (about 10 Kohm). Design of buffer input/output circuit employs same LM102H amplifier on both ends of interface for proper impedance matching between the output impedance of SCU buffer and the input impedance of ADC buffer has been used to avoid such problem.

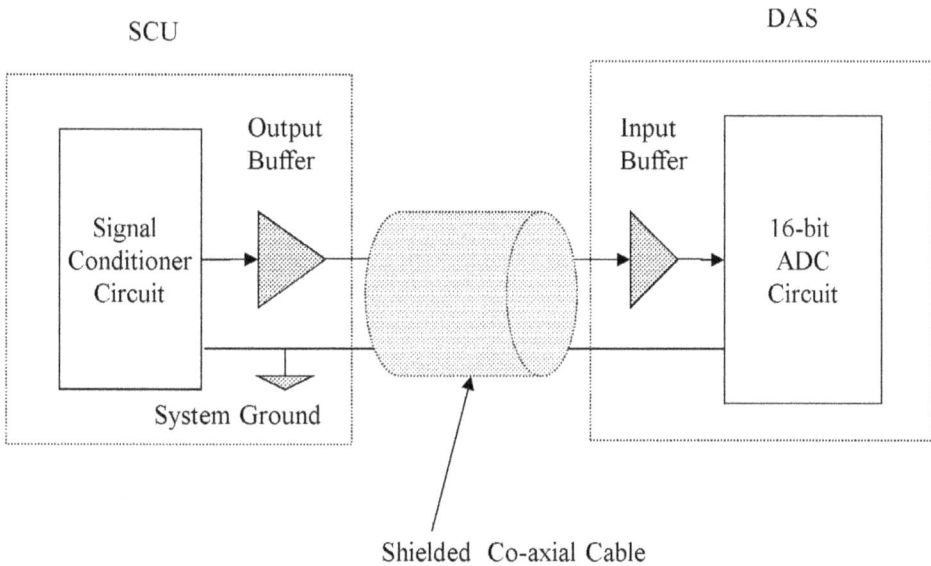

Fig [4]: *End-to-End Buffer Implementation Between Output of Signal Conditioner and ADC of data acquisition system*

The output signal as arrived at the input of the ADC through buffer amplifier is converted into a corresponding digital word of 16-bit. The ADC operates under the control of software. DAS is configurable as per the requirement of user like the number of channels, sampling interval, and dates etc. The user can select as many channels as are required. ADC reads only selected number of channels. The system block diagram is shown in the figure 5.

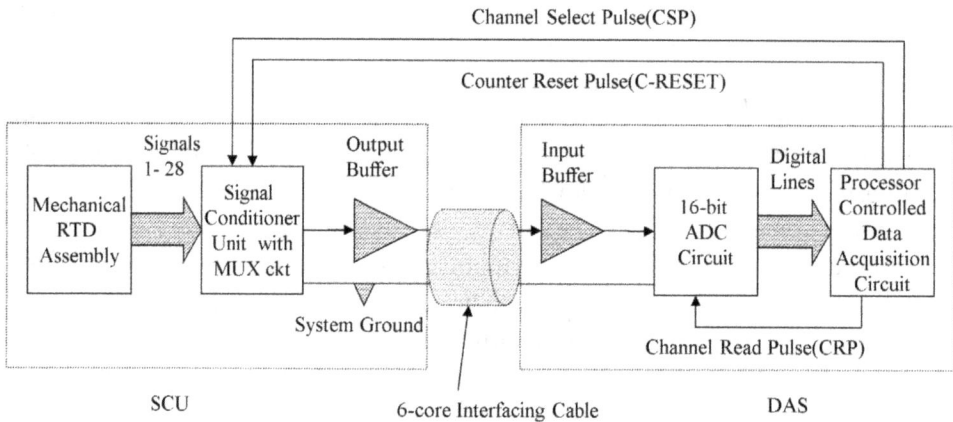

Fig [5]: *Complete Block Diagram of Interfaced Sub-systems (SCU and DAS)*

The software in conjunction with the hardware generates a stream of pulses equal to the number of channels as shown in figure 6.

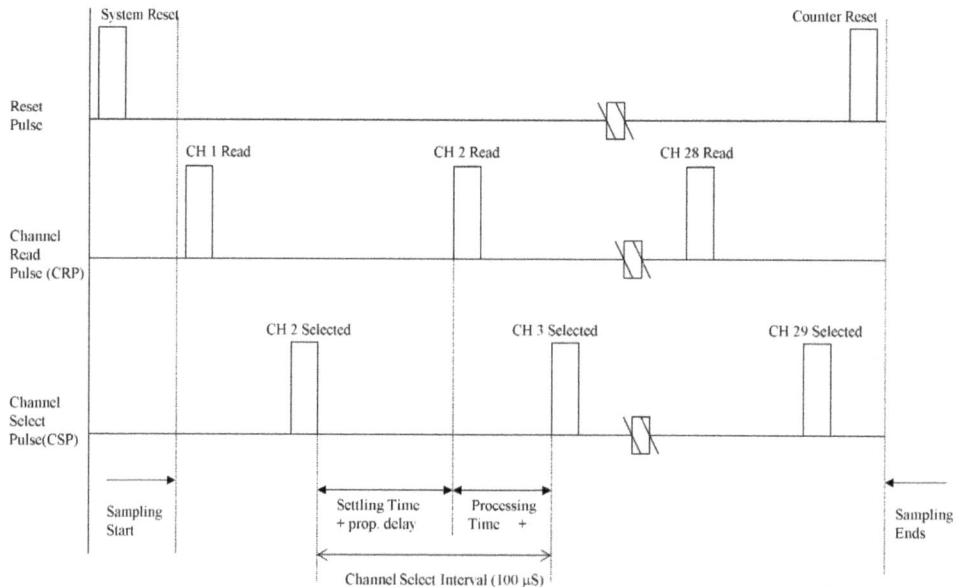

Fig [6]: *Channel Read and Channel Select Pulse Stream Pattern*

The interval between the successive pulses is dependent upon the *settling time* of the signal conditioner when multiplexers transit from one channel to another as well as *processing time* of DAS. We observed a *settling time* of 24 µs during which a particular pre-amplifier channel voltage is settled at the output of MUX and travel through the following gain circuitry and appear at the buffer output after the said channel is selected. For convenience *propagation delay time* (approx. 17 μs) taken by the output to reach at ADC input through long cable and input buffer amplifier is also added along with above time and total transit time is termed as settling time only.

When power is applied to the signal conditioner, the counter shall be at the initial stage of 000. The combination will select the channel number one (CH 1). The analog voltage of first pre-amplifier passes through the multiplexer and appears at the output of signal conditioner through the following gain circuit and then through the cable and buffer amplifier, it appears at the input of the ADC after *settling time*. Channel Read pulse (CRP), which forms the Convert Command of ADC, is generated to read the data from the ADC. ADC will convert the analog voltage present at its input the moment conversion pulse is sent. The converted digital data is stored in memory location along with the channel identification number. The *processing time* for these actions is around 27 μs as mentioned above. The channel identification comes from the number of the read pulse generated by the software.

The same Channel Read pulse (CRP) is delayed by a time equal to or more than *processing time* and is used to generate pulse termed as Channel Select Pulse (CSP), given to multiplexer circuit to select the next channel. It increments the counter to select the next channel and hence the voltage corresponding to this selected channel appears on the input of the ADC. Sufficient settling time is allowed to voltage to settle down at ADC input before reading it. After this fixed delay equal to or more than *settling time*, next read pulse (CRP) is sent to ADC for next conversion of data and again the data is stored in memory locations with auxiliary information.

Channel Select Interval = (Previous Channel Output settling time + Time taken by output
voltage to propagate to ADC input) + (Time taken to read the Input at
ADC + Software Processing Time)
= *(Settling Time + Propagation Time) + Processing time*
= (24 μ S+ 17 μS) + 27 μS
~ 68 μ S

The safe margin between adjacent channel select pulses has been kept. When a pulse is sent from the DAS, the propagation delay caused by the length of the interfacing cable has been taken into account. *Settling Time* governs the delay between channel select pulse (CSP) and channel read pulse (CRP) and also affects the minimum time interval between two channel select pulses. This interval has to be more than the settling time, as safe margin has to be provided for an extreme transition of voltage when adjacent channels have completely out of phase voltages. The data on the selected channel should be read-only after a delay equal to *settling time*.

Further, the *processing time* is added to this *settling time* to find out the minimum value of time interval between two Channel Selecting Pulses (CSP) which do not clash the intervening events. For this *Channel Select Interval*, a safe value of 100 µS has been chosen. The results have been very reliable with this time interval setting.

After the stream of pulses equal to the number of channels configured, RESET pulse is generated which makes the counter zero and the first channel gets selected again. Depending upon the sampling interval, the same stream of pulses is generated after sampling interval.

6. *GROUNDING AND SHIELDING ASPECTS IN INTERFACING*

Experimentally some output voltage drops of about 82 mv have been observed due to long cable length, even if high-quality thick cable wires have been used. The major problem encountered has been the shifting of the ground level by 82 mV due to the introduction of cable resistance in the ground loop and series-mode voltage. The buffers only eliminate the loading and the impedance reflections, but ground level shifting cannot be eliminated in this manner.

6.1 *Guarding Input Signal Pins*

As preliminary protection, the input signals to Operational Amplifiers and ADC are properly guarded using PCB layout techniques [6]. Analog input signal pins are guarded with filling thick copper track enclosing the pins and providing a thin aluminum shield as shown in figure 7.

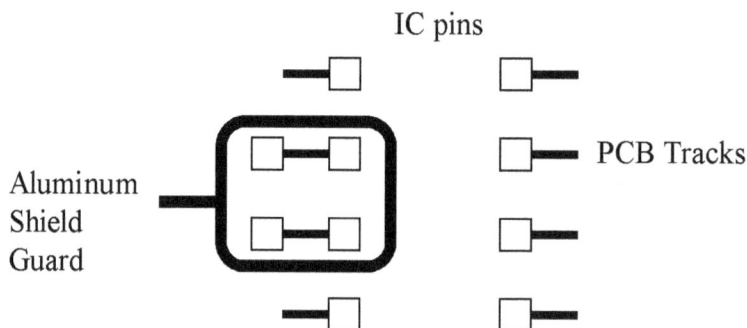

Fig [7]: *Board layout for Guarding Inputs*

This guarding is achieved by applying a low impedance bootstrap potential to the outside of the insulation material surrounding the signal lines. This bootstrap potential is held at the same level as that of the high impedance signal line, therefore, there is no voltage drop across the insulation and hence no leakage. This guard acts as a shield to reduce noise pick up and serves an additional function of reducing the effective capacitance to the input line. In some of the op-amps, the IC case is connected to guard potential. This virtually eliminates the potential leakage paths across the package insulation, provide noise shield for the sensitive circuitry and reduce common mode input capacitance to about 0.8 pF.

6.2 Separating Analog and Digital Ground at ADC

Secondary protection is a separation of Analog and Digital Ground in the circuit to minimize interference of analog and digital circuits. In order to isolate low level analog signals from a noisy digital environment, the ADC has two ground pins. These are Analog Ground (AGND) and Digital Ground (DGND). These ground pins are tied together at one point only. If a single ground line is used or AGND and DGND are connected at ADC pins itself, a current through ground wires can cause the hundreds of millivolts of error. Separate ground returns to minimize the current flow from sensitive points to the system ground. Refer to figure 8. For noise immunity, a 47-microfarad tantalum capacitor is connected between analog and digital ground at ADC.

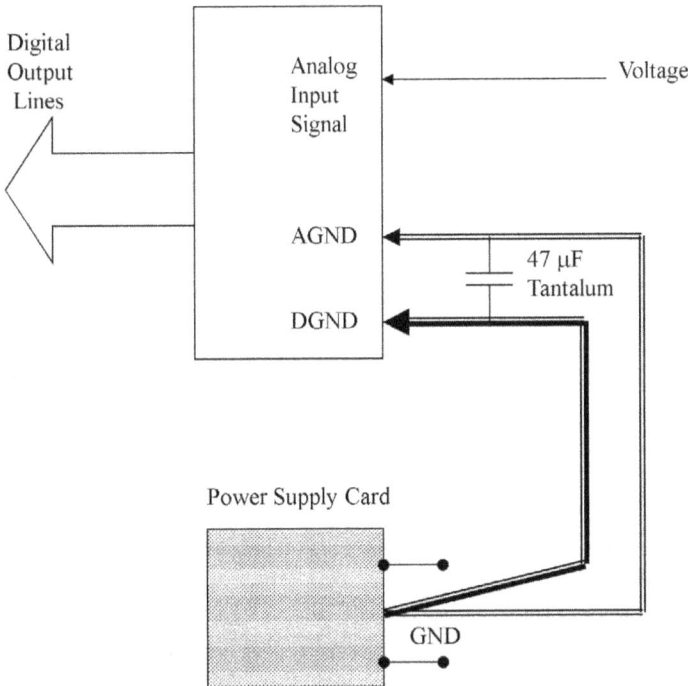

Fig [8]: *Separate AGND and DGND at ADC*

6.3   *Interconnecting Analog and Digital Ground in Multiple Cards*

Our Signal conditioner, as well as DAS, has been made in form of modular PCBs. Care has been taken in keeping separate digital and analog ground on every card and connecting them together at one point (power supply card in this case) only. The point of interconnection is the system ground which is connected to shield and earthed. This technique is depicted in figure 9.

6.3.1   *Using Ground Planes on PCBs*

The PCB layout and designing aspects also considerably improve the system performance if certain rules related to Analog and Digital Ground are taken into consideration. In a mixed signal design of Multi-point Snow Temperature Sensing system, analog and digital portions of the board are made distinct from one another with Analog ground plane covering analog signal traces and the digital ground plane confined to the areas covering digital interconnect. Care has been taken to ensure that the ground plane is uninterrupted over crucial signal paths. Wide runs of tracks or planes in the routing of power lines has been done. This served the dual function of providing a low series impedance power supply as well as free capacitive decoupling to the appropriate ground plane.

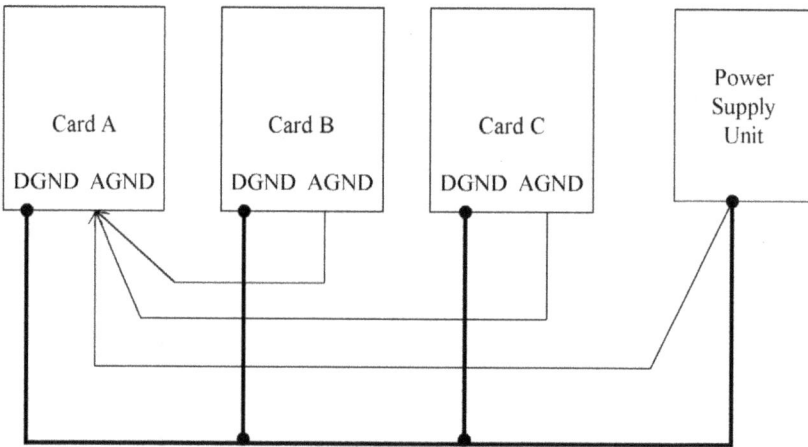

Fig [9]: *Grounding procedure for Multiple Card of Signal Conditioner*

6.3.2   *Using Power Supply Bypass Capacitor*

Power supply bypass capacitor is also used which minimize series resistance and inductance. These are installed on PCB with the shortest possible leads consistent with the reliable construction. We have used 0.01 µF CTR coupling capacitors on all the IC power supply points.

6.3.3   Shielding and Earthing of the System Cabinets

Further, proper shielding techniques have been used in the design. Cabinet shield and cable shielding is used to protect the system from stray currents and noise. The shield is only effective when the shield is connected to the common or zero reference potential of the shielded circuitry eliminating the undesired feedback path [13]. The standard practice is to connect the shield to the ground or common point of the circuit and then to earth it at a suitable point. The present system consists of two independent cabinets to be connected together. It is known from the instrumentation theory that if multiple shielded cabinets are wrongly connected, a sensitive magnetic loop is created [14]. The major design practice in shielding techniques is the selection of proper point where to earth the overall system shield.

In the present case, we consider two obvious options for earthing of the cabinet shields. First is to earth one of the SCU or DAS shields and second is to earth both the cabinet shields. Experimentally both the alternatives produced magnetic sensitive ground loops [15].

The first alternative to earth the DAS cabinet is shown in figure 10. C1 and C2 represent the parasitic capacitance through which two cabinets get coupled to the mains V_m. A current i_l flows through C1, Cabinet1, and shield of the coaxial cable to the ground connection of Cabinet 2. This forms the series-mode voltage error and will be either subtracted or added to the input signal for DAS. The error finally gets reflected in the temperature measurement.

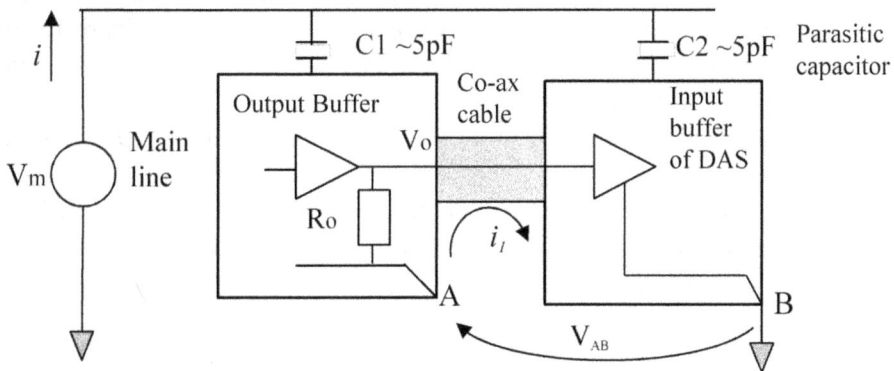

Fig [10]: Ground Loop Interference produced in an interconnecting coaxial cable when at only DAS cabinet is provided with Single point Earthing

In the second alternative, if both the cabinets are earthed at their respective ends as shown in figure 11, still a voltage difference get developed between the cabinets and a magnetically sensitive loop is created. A series-mode voltage error of the order of 90 mV has been observed in this configuration, which straightaway indicated an error of 0.9 °C in the temperature measurement.

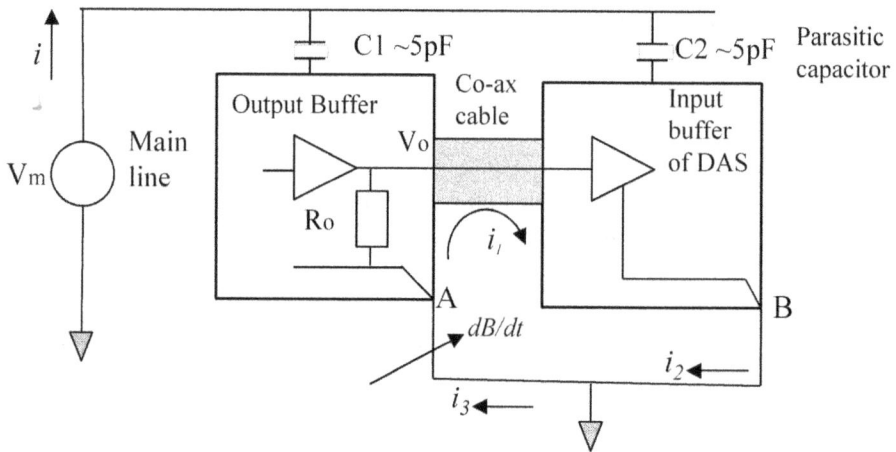

Fig [11]: *Magnetic sensitive ground loop created in an interconnecting coaxial cable by grounding both signal conditioner and DAS cabinets*

A third alternative [15] has been executed to solve this problem using two-wire shielded cable (in case of one signal only) and applying the differential input stage at the input of the shielded circuitry in the second cabinet as depicted in figure 12. In the present case, 6-wire shielded cable has been used because signals to be interchanged are more. Here Common of the entire circuit is connected to shield and earthed at one point only. This configuration gives highly improved results and minimizes the interference even if a long cable is used. No ground loop exists, and parasitic currents cannot influence any input or create a series-mode voltage. The series-mode voltage is reduced to less than 20 mV which corresponds to better than 0.2 °C accuracy in the overall system when two sub-systems are interfaced.

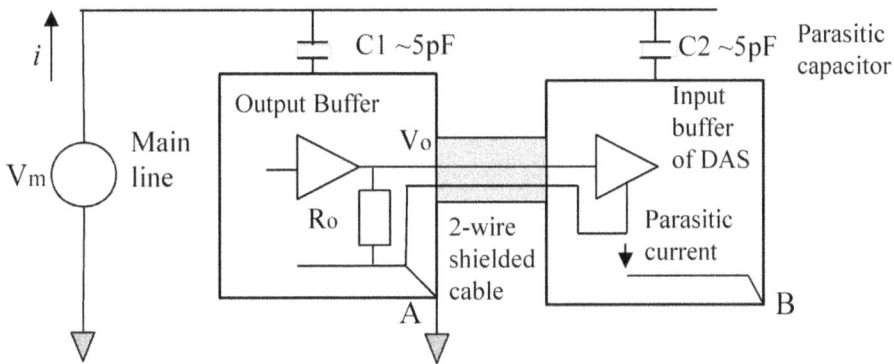

Fig [12]: *Interfacing of signal conditioner with DAS using two-wire shielded cable and grounding at signal conditioner to avoid interference of ground loop and influence of parasitic capacitors*

7. NOISE REDUCTION ASPECTS AND RELATED DESIGN CONSIDERATION

Overall interfacing has been done using the standard rules for impedance matching, isolation, shielding and guarding [12]. But the system when integrated with the measuring system in the actual environment is quite prone to noise, which is generated from multiple sources. This noise may sometime severely interfere with the system performance and impairing the measurement.

Guarding techniques along with the proper shielding techniques, as discussed in the previous section, have been employed to prevent stray currents entering sensitive electronic parts by connecting the appropriate parts of a circuit to earth or ground. Proper shielding for magnetic and capacitive coupling and isolation for current coupling from electrical lines has been ensured as discussed above in the third alternative of grounding & shielding.

The interfering signal created by external fields tends to be the common mode, so will be suppressed only if the common mode rejection ratio (CMRR) of the system is high. CMRR is given by:

$$CMRR = A_d / A_{cm}$$

Where
A_d = Differential Voltage gain of the pre-amplifier
A_{cm} = Common Mode Voltage gain of the pre-amplifier

In terms of decibels, the CMRR is

$$CMRR_{db} = 20 \log (CMRR) = 20 \log (A_d/A_{cm})$$

High CMRR amplifiers with CMRR ratio better than 110 dB have been used in order to remove the common mode signal generated due to constant noise. In the present case, we have not used any amplifier in the differential mode because the differential configuration generates 3 dB (41.4 %) more noise than Single Ended version. Selecting an amplifier with better CMRR ratio improves its rejection response to common mode interference. In terms of signal-to-noise ratio at the input SNR_{in} and output SNR_o of the amplifier:

$$CMRR = \sqrt{(SNR_o / SNR_{in})}$$
$$\text{Or} \quad SNR_o = SNR_{in} (CMRR)^2$$

If we assume signal-to-noise ratio at the final output at the cable is SNR_{fin}, then the processing gain (G_p) of the SCU defines the change in the SNR between the input and output of the subsequent circuit:

$$G_p = SNR_{fin} / SNR_o$$

Ground loop voltage rejection techniques are essential in the present case because two self-contained units with independent power supplies are being interfaced together. The independent supplies are essential, as there will be a large voltage drop in the cable if supplies are generated in DAS and sent to the signal conditioning unit. Further, Strong ground loop current flows between the two units due to their separate power supplies and may cause interference. In this case, as a standard practice cable connector is isolated from the shield of the second cabinet.

A long multi-core shielded cable has been used to connect RTD element with the SCU in the present case. This shielded coaxial cable is used for noise reduction, but it can also add problems from three sources. i) Cable Leakage, ii) noise, iii) capacitance. We have used a virgin Teflon cable to achieve the highest possible insulation. But sometimes cable systems cannot be made very rigid and chances of spurious noise signals are fair when it is subjected to mechanical vibration, flexure, or distortion in a heavy snowbound environment. Here cable movement cause noise signals of three types:

a) Frictional movement of the shield over the insulation material generates a charge, which is sensed by the signal line as noise voltage.

b) Cable movement will also make small changes in the internal cable capacitance and capacitance to other objects. Since the total charge on these capacitances cannot be changed instantly, a noise voltage results as:

$$\Delta V = Q / \Delta C$$

c) Noise voltage is also generated by the motion of a conductor in the magnetic field.

Conductor to shield capacitance of coaxial cable is about 30 pF/foot. Charging this capacitance can cause considerable stretching of signal rise time and hence the settling time. Even the benefits of using low input capacitance amplifier are offset by this phenomenon. To solve this problem, in the pre-amplifier section, integrating amplifiers configuration (with capacitors in the feedback) has been employed to smoothen the effect of this uncontrollable noise. Refer to figure 1, where a value of C_f equal to 10 μF has been used in parallel to a resistor value of 330 K ohms. It gives a betterment of S/N ration in the system response. The amplifier is an integrator configuration, which averages the noise and acts as the low pass filter for the noise with the time constant RC seconds. This capacitor further reduces the effect of shunt capacitance at the input of the amplifier and avoid ringing in the amplifier response.

This process of putting a capacitor across the resistance is called Bandwidth reduction. For signal, the bandwidth is taken to be -3 dB cut-off frequency, where the ratio of output to input is 70.7 % or mean square power ratio is 50 %.

$$\left| e_o^2 / e_i^2 \right|_{fc} = 0.5$$

where f_c of the circuit is given by $1/(2 \Pi RC)$.

The effective equivalent noise bandwidth is given by:

$$B_n = (1/G^2) \int_o^\infty |H(j\,\omega)|2 \, df$$

Where

H $(j\,\omega)$ = frequency response function of the system

G = Gain parameter suitably chosen to be a measure of the response of the system to some parameter of the signal

For low pass filter, G is usually taken to be Zero-frequency gain, which is unity. We get

$$B_n = 1/(4RC)$$

Output noise is given by

$$E_n = e_n \sqrt{B_n}$$

Where e_n = output noise spectral density

The stray capacitance of the order of 2.5 pf sustains in the amplifier circuits.

$RC = 330 \times 10^3 \times 2.5 \times 10^{-12} = 0.825\ \mu S$

and $B_n = 303$ KHz

and output noise voltage is E_n will therefore be:

$E_n = e_n \sqrt{B_n} = 17.4\ e_n$

The signal voltage is E_s is given by

$E_s = i_s . R = 0.01 \times 10^{-6} \times 330 \times 10^3 = 3.3\ m\ V$

The Signal-to-noise ratio with parasitic capacitor only is

$SNR_1 = E_s / E_n = 3.3 \times 10^{-3} / (17.4\ e_n) = 18.5 \times 10^{-5}/e_n$

Now the capacitor of value 10 μF is put across the resistance, taking effective capacitance of 10 μ F, the noise bandwidth is

$B_n = 0.075$ Hz

And noise voltage now reduces to

$E_n = e_n \sqrt{B_n} = 0.275\ e_n$

The revised signal-to-noise ratio after adding the capacitor is

$SNR_2 = 3.3 \times 10^{-3} / 0.275\ e_n = 0.011/e_n$

So, the signal-to-noise ratio improvement is

$SNIR = (S_o / N_o) / (S_i / N_i) = \sqrt{B_n}/\sqrt{B_{no}} = SNR_2 / SNR_1 = 59.7$

This SNIR is calculated for capacitor only. In actual practice, other interference noise such as interfacing cable noise and power supply noise causes the SNIR ratio well below 30, which is still satisfactory improvement in the present case.

The long interfacing cable caused a more practical problem in the present case. It has been found that if propagation delay is equal to or greater than the pulse transition times, the effect of reflections in the cable due to termination mismatch may occur. As the digital signals (selection pulses) are being passed through this cable, overshoot and undershoot and ringing may occur along with crosstalk between the lines, the magnitude of which well exceed the noise margin of digital circuitry. Proper impedance matching as explained in previous sections has to be considered critically otherwise it may cause wrong selection of channels at the multiplexer (due to overshooting of voltage well above noise margin at maximum threshold for low level) in the worse case, and none of the channels is selected in worst case (due to undershoot, voltage going negative below the minimum threshold of high level of digital ckt. The propagation delay through the cable is kept quite low by avoiding unnecessary capacitors at buffer stages and using low capacitance interface cable.

The noise margin of the overall system has been reduced well below 20 mV from around 200 mV with combined S/N ratio improvement of more than 30 times. Overall system accuracy has been found to be 0.2 °C, when total system is integrated together.

8. CONCLUSION

The multi-point measurement system has been specifically used in the snow hydrological studies which implies a very tough environment and hence impose tight design specifications. The reliability of data is absolutely critical in such forecasting applications. The system has been designed as per the J55555 specifications with the components conforming to 883-B/JM38510 specification code. Most of the problems were practical, not having much theoretical significance and have been solved with a long experience of working team in research, design, and development of instrumentation. The design aspects and techniques for effective and reliable interfacing of Multi-point temperature measurement System with data acquisition system discussed here can be extended to any other general measuring systems placed at a long distance from the data acquisition system. Further same techniques can be applied for multiplexing of same or different kind of sensor to a common data collection platform.

REFERENCES

[1]   Shamshi, MA, Attri, RK, Sharma, VP, 1996, Snow Pack Temperature sensor, Proceedings of National Conference on Sensors and Transducers, pp. 180-189.

[2]   Singh AK,1994, Mathematical model for study of temp profile within Snow Cover, Proceedings of SNOWSYPM-94, pp. 49-52.

[3]   Satyawali PK, 1994, Grain Growth under temp gradient, Proceedings of SNOWSYPM-94, pp 5-8.

[4]   Ganju A, 1994, Snow cover model, Proceedings of SNOWSYPM-94, pp. 221-226.

[5]   Bader, HP and Wielenmann P, 1992, Modeling temperature distribution, energy and mass flow in snow-pack. Cold Region Science and technology, Vol 20, pp. 157-181.

[6]   Application Notes: Practical Temperature Measurements, Hewlett-Packard.

[7]   Rosemount, 1962, Platinum Resistance Temperature Sensors, Bulletin 9612, Rosemount Engg Co.

[8]   Temperature Measurement Handbook, Omega Engg Inc, Stanford, Connecticut.

[9]   JEvans, JP and Burns, GW, 1962, A study of stability of High temperature platinum Resistance Thermometer (Temperature-its measurements and Control in Science and Industry), Reinhold, NY.

[10] Platinum RTD Resistance table, Omega Temperature Catalogue, Vol 26, Omega Engg Inc.

[11] Tucker, JM, 1991, Fundamentals: Connecting to your Sensor, Sensor Review, Vol 11, No1, , pp. 24-27.

[12] Arnold, ADSE, 1981, Principles of Electronics Instrumentation, Prentice Hall, London.

[13] Norton, HR, Handbook of Transducers for Electronics Measuring Systems, Prentice-Hall, New Jersey.

[14] MunRoe, DM, 1982, Signal To Noise Ratio Improvement, Handbook of Measurement Science Vol. 1, John Wiley Sons.

[15] Morrison, R, 1986, Grounding and Shielding Techniques in Instrumentation, John Wiley & Sons, California.

[16] Attri, RK 2018/1996, 'Snow Pack Temperature Profile Sensor,' R.Attri Instrumentation Design Series (Snow Hydrology), Paper No. 4, *Research and Design of Snow Hydrology Sensors and Instrumentation*, 2nd edn., pp. 59-64, Speed To Proficiency Research: S2Pro©, Singapore.

[17] Attri, RK, 2018/1999, 'Design of an Instrumentation System to Record Distribution Profile of Snow Layer Temperature for Modelling of Snow Avalanche Forecast,' R.Attri Instrumentation Design Series (Snow Hydrology), Paper No. 5, Research and Design of Snow Hydrology Sensors and Instrumentation, 2nd edn., pp. 66-75, Speed To Proficiency Research: S2Pro©, Singapore.

[18] Attri, RK, 2018/2000, 'Design Approach to Use Platinum RTD Sensor in Snow Temperature Measurements,' R.Attri Instrumentation Design Series (Snow Hydrology), Paper No. 6, Research and Design of Snow Hydrology Sensors and Instrumentation, 2nd edn., pp. 66-75, Speed To Proficiency Research: S2Pro©, Singapore.

[19] Attri, RK, 2018/2000, 'Design of A True Snow Air Temperature Sensing Probe,' R.Attri Instrumentation Design Series (Snow Hydrology), Paper No. 8, Research and Design of Snow Hydrology Sensors and Instrumentation, 2nd edn., pp. 66-75, Speed To Proficiency Research: S2Pro©, Singapore.

[20] Shamshi, MA, Attri, RK & Sharma, VP, 1996, "Snow pack temperature sensor," Proceedings of National Conference on Sensors and Transducers, Chandigarh, pp. 180-189, viewed 24 Jan 2018, <https://www.researchgate.net/publication/275276742>.

[21] Attri, RK, Sharma, BK, Shamshi, MA & Sharma VP, 2000, 'Design Approach to use Platinum RTD Sensor in Snow Temperature Measurements', Journal of Instruments Society of India, vol. 30, no. 4, pp. 275-283, available at https://www.researchgate.net/publication/275276709.

[22] Attri, RK, Sharma, BK & Shamshi, MA, 2000, 'Practical Design Considerations for Signal Conditioning Unit Interfaced with Multi-point Snow Temperature Recording System', IETE Technical Review, vol. 17, no.64, pp. 351-361, https://doi.org/10.1080/02564602.2000.11416928 or download from https://www.researchgate.net/publication/275276698.

Paper No.8

DESIGN OF A TRUE SNOW AIR TEMPERATURE SENSING PROBE

Raman K. Attri
Ex-Scientist,
Central Scientific Instruments Organization INDIA

Abstract — *One of the key temperature parameters in snow temperature profile system is true ambient air temperature. Measuring true air temperature due to factors like wind and solar radiations is a challenging task. In this paper, a design of snow air temperature sensing probe is described. The paper also discusses the design techniques to address various accuracy, precision and linearity errors encountered in the temperature measurement system.*

1. INTRODUCTION

Some of the phenomena such as climatic changes, the potential release of avalanches, river run-off water, glacier sliding etc. in the mountain areas and planes nearby are closely associated with snow. To carry out forecasts for this phenomenon, snow parameters have to be measured critically. Temperature has been identified as the most important parameter for snow hydrological studies. Snow is a thermodynamically unstable material, which undergoes morphological changes. The formations of snow cover take place with the development of different forms of crystals. The shape, size, bonding, and packing of these crystal controls the mechanical properties and hence stability and strength of snow cover thus evolved. The stability, strength, and structure of different layers composing snow pack mainly depends upon temperature distributions within and outside snowpack. Temperature affects the net exchange in at snow-air interface, snow-ground-interface, and snowpack itself. The energy at the snow-air interface is considered to deduce the surface melt, energy exchange at the snow-ground interface examines the possibility of destruction of the bond between the ground and the snowpack.

Among all, true air interface temperature is one of the highly important parameters to be measured for snow hydrological studies. So, air temperature above snow surface is needed to be measured critically by instrumentation systems. The snow-air temperature sensor measures the true air temperature (free from the effect of solar radiation, sunlight, etc.) This sensor is used to get the hourly value of average Air Temperature. This physical monitored temperature data from this sensor is used to provide extremely useful information on temperature distribution outside the snowpack. This data is used in the computation of melt, crust thickness, and run-off water. The computed parameters are further used in a computer-aided simulation model of snow cover formation, thickness, melt, and strength. The above computations are incorporated in snow avalanche forecast modeling, climatic forecast modeling and river run-off water determination.

The energy distributions inferred from the temperature distribution data are used as follows:

- To evaluate the snow-melt from snow-air energy exchange.
- To develop snow-melt run-off models of rivers and a net rise in river water level. This data further leads to flooding forecast modeling.
- To study the possibility of melt and destruction of the bond between ground and snowpack. This leads to forecast of formation and release of avalanche and glacier sliding etc.
- To study the process of metamorphism of snow for evaluating the strength and stability of snowpack.
- To forecast the climatic changes due to changes in snow cover and snow melting.

2. DESIGN REQUIREMENTS OF THE SYSTEM

The system for air temperature measurement in snowbound areas requires great design efforts, and tight specifications are imposed upon it. The harsh environment imposes great design restrictions and tight specifications upon this system and great performance reliability. As the system has to operate round the clock in the harshest possible environment in heavily snowbound areas, it requires extra environmental raggedness besides electronics circuit reliability. The probe has to operate round the clock for several months under the following conditions: -

i.   Temperature range of operating is -50°C to +50°C.
ii.   Relative humidity to be handled at the place of measurement is likely to be 100%.
iii.   Wind speed of the order of 200 Km/h is to be handled.
iv.   Moisture ingression in the system has to nil.
v.   The system should be unaffected by rain or snowfall.
vi.   High rigidity, strength and tight packaging of the system.
vii.   Accuracy in temperature measurement should be 0.1°C.
viii.   Resolution of temperature sensing probes is to be 0.1°C.
ix.   Current consumption to be less than 100 mA.

Given these severe environmental conditions, the packaging has to be water and moisture proof & wires and cables are to be selected so that these can withstand these operational conditions. The performance of the design has been very reliable and up to the accuracy of 0.1°C. Because of actual requirement, the minimum use of components, weight, current consumption, and cable lengths, etc has to be considered in the final design. The snow air temperature sensor utilized J55555 specifications with the components conforming 883-B/JM38510 specification code.

3. SYSTEM DESIGN

The snow air temperature sensing probe is usually mounted on the top of a mast, installed in the heavily snow bound areas. This mast is fitted with several other temperature sensors meant for measuring temperature at different points, including snowpack and ground temperature. A typical installation arrangement is shown in the figure 1. A cable connects this sensor to the signal conditioner unit, which is a weatherproof compact unit.

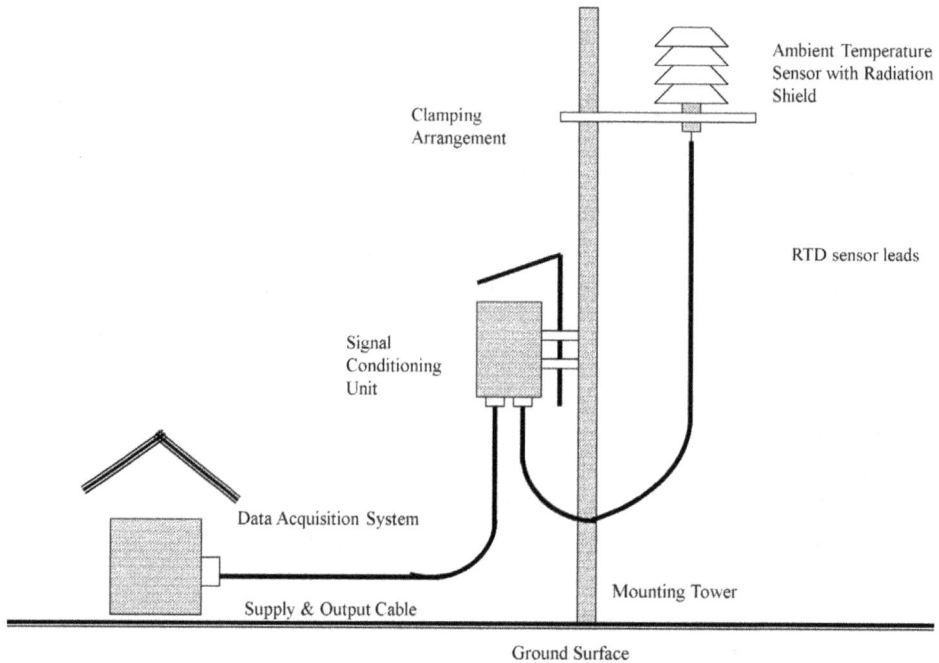

Fig [1]: *Mounting of snow air temperature sensing probe*

This signal conditioning unit houses the electronics, the signal from which is then routed to a data acquisition system through a long low-temperature cable. Figure 2 shows the block diagram of the snow-air temperature sensor. Data acquisition system, which is a processor-based unit with self-contained memory modules. The reading of temperature sensor is taken after selectable timing interval.

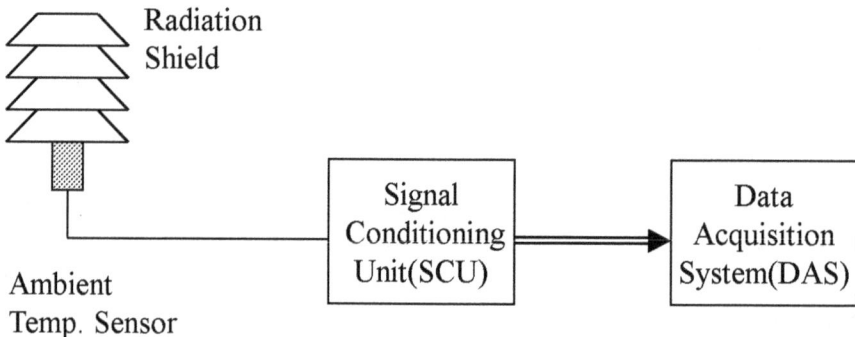

Fig [2]: *Block diagram of snow air temperature sensing probe*

On-chip software computes minima, maxima & average temperature of the air and stored with time stamps. Data can be retrieved when required from the data acquisition system via RS-232 Interface of PC. Software on the PC processes the data and plots temperature vs. time. This gives air temperature variations with time.

3.1   Fast responding temperature sensor that reduces self-heating

Out of commonly used sensors viz. RTD, Thermistor, Thermocouple, the design factors and specifications like sensitivity, accuracy, linearity, error compensations and environmental conditions have been taken into consideration to select the most suitable temperature sensor. Thermistor's resistance versus temperature characteristics are logarithmic and is not uniform over the entire range of required measurement, making it is more non-linear than RTD. Thermocouple does not work satisfactorily because metals need to be protected in moisture environment. Cold junction compensation, reference junction temperature control, and lead compensation makes the measuring unit quite bulky and complex. Given this, Platinum RTD with PT-100 element has been used for this application. Metal Film RTD sensing element has been used in the sensor assembly. The films RTD have the advantage of increased resistance for a given size. Due to small size, it can respond quickly to step changes in temperature, and the measurements are decidedly improved. Some of the salient points regarding PT-100 RTD sensor are high precision and accuracy, ease of calibration, high repeatability & fast response, interchangeability with other resistance thermometers without any compensation, good performance in desired temperature range and limited susceptibility to contamination, etc.

The RTD sensing element is sealed into a metallic sheath of steal which is a hollow metallic housing that contains the sensing element and rest of the space is filled with a thermal conductor aluminum oxide to reduce self-heating. This self-heating is a result of Joule heating caused by current passing through RTD. The sensing assembly is shown in figure 3. The heat produced due to self-heating is instantly absorbed by aluminum oxide preventing a rise in temperature of the platinum element.

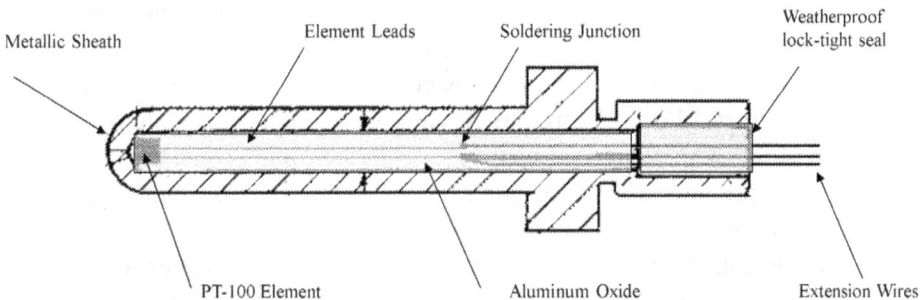

Fig [3]: *PT-100 sending element enclosed in Aluminium Oxide*

3.2   Design for true air temperature measurement

Snow-air temperature sensor consists of one RTD sensor element which is enclosed in a self-aspirated shield so that sensor gives the true air temperature and is not affected by solar radiation. The radiation shield is stacked plates fastened together, as shown in figure 4.

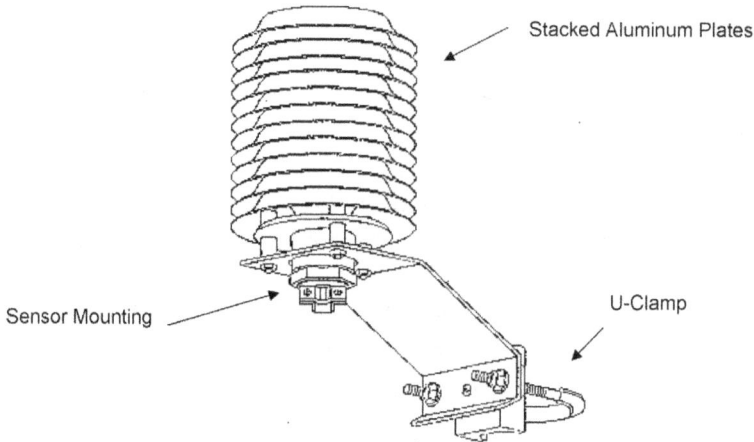

Fig [4]: *Snow air temperature sensor shield*

The air flows through the ventilation of this shield and raises the temperature of the sensing element. The radiation shield is coated with special UV film for protection from sunlight. Thus, the heating caused by solar radiation and cooling caused by snowfall and rain does not affect the reading of this temperature sensor. Only the true air temperature is read.

3.2   Amplifier That Is Sensitive to Small Changes in Temperature

The nominal resistance of PT-100 element at 0^oC is 100 ohms. The platinum RTD has a positive temperature coefficient of resistance variation. With the increase in temperature, the resistance of RTD increases at a rate of 0.39 ohm per degree Celsius change in temperature. The circuit used to detect such a minute change in resistance has to be very sensitive. The design of such sensitive electronics consists of a reference voltage section, pre-amplifier section, gain amplifier section and non-linearity compensation section.

The reference voltage module uses a stabilized reference voltage generator with an average temperature coefficient of 0.002 %/^{o}C in the range of -55 ^{o}C to +125 ^{o}C and the long-term stability is 0.1%/1000 hrs. Using military specification amplifier and thin film 0.1% resistance and Stereoflex/Tantalum capacitors, a highly stable reference voltage of 2.5V is obtained. This reference voltage input is fed to pre-amplifier that converts changes in RTD resistance due to temperature variations into corresponding voltage readings, as shown pre-amplifier section in figure 5.

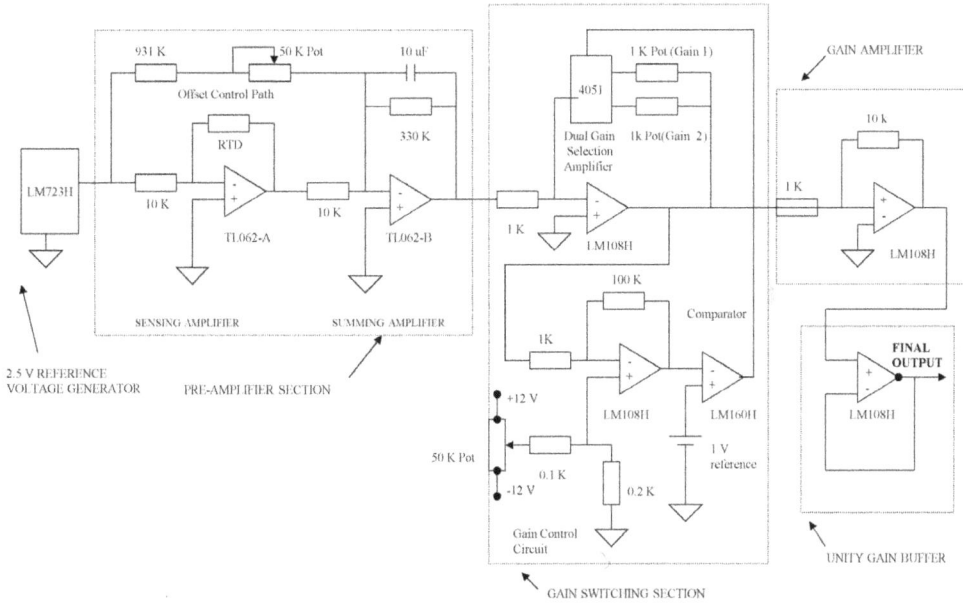

Fig [5]: *Circuit Diagram of snow air temperature sensing probe*

The pre-amplifier is a dual operational amplifier which uses RTD in the feedback path of the first amplifier. By designing in this manner, the current passing through RTD has been of the order of microamperes, thereby reducing the heating almost to a negligible value. Error due to self-heating is reduced by passing the minimum ohms measurement current that gives a desired resolution. The voltage proportional to the change in resistance of RTD is obtained at the output of the first amplifier which is then amplified by a factor of 3.3 by a gain amplifier.

Now at 0°C, the resistance of RTD is 100 ohms, the offset control (a variable resistance) is adjusted in such a way that makes output voltage to read zero corresponding to zero degree Celsius. 100-mv corresponding to 1°C change in temperature is obtained at the output of the final gain amplifier.

3.3　*Offset control circuit that compensates errors due to the extension leads*

The offset control circuit shown with a feedback loop and a variable resistor in figure 5] in the pre-amplifier section also helps to remove the error caused by long extension leads which may run to several meters. A severe error is caused even if high-quality extension wires of 1 ohm/4m resistance are used. Offset control compensates the error caused by lead length by injecting an equivalent voltage in the summing amplifier.

The change in resistance of the extension wires due to temperature also needs to be compensated. To nullify this error, three wires RTDs has been used in the feedback path. The design configuration evolved in this way enables the system to respond only to resistance variation in RTD and eliminates lead error in conjunction with offset control.

3.4   Gain Switching Mechanism That Compensates for Non-linearity

The linearity of the sensing element, as well as its fast response to minute changes in temperature, is an important design consideration. Platinum sensing element exhibits some non-linearity also, which is compensated by electronics circuit. From resistance-temperature curve of platinum in the range of -50°C to +50°C, it is inferred that the dispersion of resistance is not equal for equal dispersion in temperature on both sides around 0°C, as summarized in table 1.

Temp in °C	Resistance (ohm)	difference
-50	80.25	0.40
-49	81.65	0.40
-48	82.05	0.39
0	100.00	0.39
+48	118.62	0.39
+49	119.01	0.38
+50	119.40	0.39

Table 1: Excerpt from Resistance-Temperature Table (European Curve, Alpha=0.00385) of Platinum RTD

The resistance of platinum at -50°C is 80.25 ohm which is a dispersion of 19.75 ohms from the zero-degree value of resistance (100 ohms). Whereas on the other side at +50°C, the resistance value is 119.40 ohm, which is a dispersion of 19.40 ohm from zero temperature value. Therefore, a net deflection of 0.35-ohm (19.75-19.40) in platinum resistance curve is obtained in the range of -50°C to +50°C temperature from linear ideal curve. The temperature coefficient of platinum is 0.39 ohm/°C and the above deflection of 0.35 ohm indicates a net error of 0.9°C in platinum temperature response on either side of the range of -50°C to +50°C.

To compensate for this error, a gain control circuit, shown as gain switching section in figure 5, is used to select a different gain in the positive and negative ranges. This is achieved with the help of proper switching at 10°C from non-linearity has been observed to begin in the positive range.

The comparator in the gain switching section compares the voltage from previous stage to a threshold of 1.0 volts which represents a 10°C point. As soon as comparator output changes polarity, it triggers a gain select pulse which is basically a square pulse to the multiplexer to select different gain resistors. That allows feedback circuit to force a different gain compensated for non-linearity difference. That means, from the range − 50°C to +10°C, a different gain is enforced through multiplexer while for the range +10°C to +50°C, a different gain is triggered. For this arrangement to work effectively, the variable resistors controlling gains in two ranges need to be calibrated at − 0°C point, - 50°C point and +50°C point and +10°C reference points to read the corresponding voltage accurately at the final output amplifier. For instance, at +50°C point, the positive gain is adjusted to get exact +5 V at the final output.

3.5   *Output amplifier that compensates for interface loading errors*

The final stage of the snow-air temperature measuring unit is the high input impedance buffer amplifier with a unity gain which follows a gain amplifier with gain of 10, as shown in the figure 5. This serves as the isolation shield between the signal conditioning circuit and the data acquisition system and further performs the function of impedance matching.

The input stage in the data acquisition system is a high input impedance buffer amplifier and it performs the same function as the output buffer at the signal conditioning system. A proper driving amplifier is to be selected which could respond well to rapid changes in input current as the analog to digital conversion proceeds. These two buffers in combination minimize the impedance reflections on both sides of the interface. It also minimizes loading of analog output by the analog to digital converter. The output signal is then converted into a digital value and stored in the memory.

4. *DESIGN TRADEOFFS AND TECHNIQUES TO IMPROVE SYSTEM PERFORMANCE*

4.1   *Inherent Errors in Platinum RTD*

In spite of the merits of platinum RTD sensor, it has serious inherent shortcomings - i) Self-heating error, ii) Extension leads error, iii) Thermal EMF error, iv) Non-linearity error, and v) Reference voltage variations error with temperature. The design needs to make use of shields and twisted-pair wire, use of proper sheathing, use of guarded integrating DVM, and avoidance of stress and steep gradients etc have also been used.

4.2   *Size of RTD*

RTD is somewhat more fragile than the thermocouple and precautions must be taken to protect it. For that, the sensing element is sealed in metallic sheath filled with a thermally conductive material like Aluminum Oxide. For reducing the self-heating, both the size and the current as well has to be considered. Minimum ohm measurement current and larger RTD (larger size gives a better reduction in self-heating), which can still give good response time have to be selected. Care has been taken not to increase the size of the sheath too much just in order to reduce self-heating.

4.3   Grounding and Shielding in Interfacing

The major problem is the shifting of the ground level due to the introduction of cable resistance in the ground loop. The buffers only eliminate the loading and the impedance reflections, but ground level shifting cannot be eliminated in this manner. Proper guarding and shielding techniques have to be used in the design. The overall system has to be properly shielded to avoid stray noise and couplings. Since the system consists of two shielded boxes, signal conditioning unit and data acquisition system, these boxes have to be properly connected together and earthed at one point only taking into view the error mentioned above. Since in present case a long coaxial cable has been used which causes considerable voltage difference, this configuration gives highly improved results and minimizes the interference.

4.4   Separation of Analog & Digital Ground

The measuring system, both the signal conditioner as well as the data acquisition system has been made modular, consisting of several boards. Care must be taken in keeping the digital and analog ground separate and connecting them together at one point only. The point of interconnection must be connected to shield and made earthed. For noise immunity a 47-microfarad tantalum capacitor connected between analog and digital ground at ADC. As a standard practice, these two grounds have been interconnected at power supply card only.

4.5   Noise Reduction

Overall interfacing requires using the standard rules for impedance matching, isolation, shielding and guarding. Still, the system when integrated with the measuring system in the actual environment will be quite prone to noise, which is generated from multiple sources. This noise may sometime severely interfere with the system performance and impairing the measurement. Guarding techniques along with the proper shielding techniques must be employed to prevent stray currents entering sensitive electronic parts by connecting the appropriate parts of a circuit to earth or ground. High CMRR amplifier has been used in order to remove the common mode signal generated due to constant noise. Proper shielding for magnetic and capacitive coupling and isolation for current coupling from electrical lines has to be ensured.

5. CONCLUSION

The snow air temperature sensing probe has been a challenging task due to factors like an exposed environment, solar radiations, the distance of sensing elements from measurement units to name a few. The key to a successful design of snow air temperature sensing probe is to incorporate critical compensating circuitry used in the analog signal conditioning unit which has led to highly stable and accurate air temperature measurement system.

REFERENCES

[1] Attri, RK 2018/1996, 'Snow Pack Temperature Profile Sensor,' R.Attri Instrumentation Design Series (Snow Hydrology), Paper No. 4, *Research and Design of Snow Hydrology Sensors and Instrumentation*, 2nd edn., pp. 59-64, Speed To Proficiency Research: S2Pro©, Singapore.

[2] Attri, RK 2018/1999, 'Design of an Instrumentation System to Record Distribution Profile of Snow Layer Temperature for Modelling of Snow Avalanche Forecast,' R.Attri Instrumentation Design Series (Snow Hydrology), Paper No. 5, Research and Design of Snow Hydrology Sensors and Instrumentation, 2nd edn., pp. 66-75, Speed To Proficiency Research: S2Pro©, Singapore.

[3] Attri, RK 2018/2000, 'Design Approach to Use Platinum RTD Sensor in Snow Temperature Measurements,' R.Attri Instrumentation Design Series (Snow Hydrology), Paper No. 6, Research and Design of Snow Hydrology Sensors and Instrumentation, 2nd edn., pp. 66-75, Speed To Proficiency Research: S2Pro©, Singapore.

[4] Attri, RK 2018/2000, 'Practical Design Considerations for Signal Conditioning Unit Interfaced with Multi-point Snow Temperature Recording System,' R.Attri Instrumentation Design Series (Snow Hydrology), Paper No. 7, Research and Design of Snow Hydrology Sensors and Instrumentation, 2nd edn., pp. 66-75, Speed To Proficiency Research: S2Pro©, Singapore.

[5] Shamshi, MA, Attri, RK & Sharma, VP 1996, "Snow pack temperature sensor," Proceedings of National Conference on Sensors and Transducers, Chandigarh, pp. 180-189, viewed 24 Jan 2018, <https://www.researchgate.net/publication/275276742>.

[6] Attri, RK, Sharma, BK, Shamshi, MA & Sharma VP, 2000, 'Design Approach to use Platinum RTD Sensor in Snow Temperature Measurements', Journal of Instruments Society of India, vol. 30, no. 4, pp. 275-283, available at https://www.researchgate.net/publication/275276709.

[7] Attri, RK, Sharma, BK & Shamshi, MA, 2000, 'Practical Design Considerations for Signal Conditioning Unit Interfaced with Multi-point Snow Temperature Recording System', IETE Technical Review, vol. 17, no.64, pp. 351-361, https://doi.org/10.1080/02564602.2000.11416928 or download from https://www.researchgate.net/publication/275276698.

INDEX

Speed To Proficiency
RESEARCH

Accelerated Performance for Accelerated Times

Highly-specialized know-how, learning, and resources to solve challenges of 'time' and 'speed' in performance at organizational, professional and personal levels.

Visit us at https://www.speedtoproficiency.com/

S2Pro© Speed To Proficiency Research is a corporate research and consulting forum that provides authentic guidelines to business practitioners to accelerate proficiency of their workforce, teams, and professionals at the 'speed of business'. S2Pro© publishes reports, ebooks, and articles exclusively related to accelerated performance, accelerated proficiency and accelerated expertise in individual and organizational context. Our extensive knowledge base of "how to methods" is derived from experience-based and practice-based observations, analysis/synthesis of existing research, or based on planned/focused research studies through a network of researchers who exclusively focus on 'time' and 'speed' metrics in the business context.

Speed To Proficiency Research: S2Pro©
A research and consulting forum
Singapore 560463

Website: https://www.speedtoproficiency.com
e-mail: rkattri@speedtoproficiency.com
Facebook: https://www.facebook.com/speedtoproficiency/
LinkedIn: https://www.linkedin.com/company/speedtoproficiency/
Twitter: https://www.twitter.com/speed2expertise
Google+: https://plus.google.com/101561704929830160312

www.ingramcontent.com/pod-product-compliance
Lightning Source LLC
Chambersburg PA
CBHW081817200326
41597CB00023B/4278